Education in a Competitive and Globalizing World

Technology in a Globalizing World

nova
science publishers
www.novapublishers.com

Education in a Competitive and Globalizing World

Life Skills in Contemporary Education Systems: Exploring Dimensions
Pooja Gupta, PhD (Editor)
2023. ISBN: 979-8-88697-833-9 (Softcover)
2023. ISBN: 979 8 88697 902-2 (eBook)

Effective Pedagogical Skills for a 21st Century Teacher: Reflections, Theories and Practices
Lawrence Meda (Editor)
2023. ISBN: 979-8-88697-919-0 (Softcover)
2023. ISBN: 979-8-89113-146-0 (eBook)

More information about this series can be found at
https://novapublishers.com/product-category/series/education-in-a-competitive-and-globalizing-world/

Technology in a Globalizing World

Management in the Digital Era: Different Perspectives
Serafettin Erten, PhD (Editor)
Mehmed Zahid Cogenli, PhD (Editor)
2023. ISBN: 979-8-88697-949-7 (Softcover)
2023. ISBN: 979-8-89113-073-9 (eBook)

The Fundamentals of Infrared Thermal Imaging
James K. Kubiak (Editor)
2023. ISBN: 979-8-89113-163-7 (eBook)

More information about this series can be found at
https://novapublishers.com/product-category/series/technology-in-a-globalizing-world/

Priya Jindal
Amandeep Singh
and Rashmi Aggarwal
Editors

STEM

A Multi-Disciplinary Approach to Integrate Pedagogies, Inculcate Innovations and Connections

nova
science publishers
www.novapublishers.com

https://doi.org/10.52305/UXLT2425

NOTICE TO THE READER

Library of Congress Cataloging-in-Publication Data

ISBN: 979-8-89113-305-1

Published by Nova Science Publishers, Inc. † New York

Contents

Preface

The final product of the education system is not able to prove its worth when gauged on the basis of practical know-how. The skill gap is a major contributor here. The need of the hour is to introduce pivotal changes in the age-old classroom practices. A dynamic system is a must to transform the world by imparting wisdom. Developing Curiosity, Innovation, and Resourcefulness might offer the solution. Curiosity and innovation should reverberate in the modern education setup. The performance should also be evaluated on the parameters of scientific literacy. The mindset and aptitude of the students need to be broadened by integration of pedagogy at two levels such as across the globe and also across the streams. Wicked problems posing challenges are coaxial to innovation. Out-of-the-box thinking by applying a scientific approach can offer relief to mankind.

The need is to build a world where denizens can understand and solve complex challenges. To achieve the demands of the dynamic and evolving workforce and to build students' skills and content knowledge literacy in STEM fields is essential. Stem's curriculum ensures that learners have access to a quality learning environment. STEM Mindset prepares learners to amalgamate scientific knowledge and skills to solve problems in a systematic manner. Stem engineering enables young minds to make sense of information and provides know-how to gather and evaluate evidence to make decisions.

The term "STEM" stands for science, technology, engineering, and mathematics. It is used to group academic disciplines. The term is used to refer to an education policy or curriculum choices in education. STEM is a unique educational curriculum that engages with multi-disciplinary pedagogical modes. It helps academicians and policymakers to bridge the existing gap. It enables curriculum developers to understand the needs of the future and enables them to produce dynamic Netizens who ace the core scientific skills to lay down the foundation of scientific solutions. With this focus in mind, the book is proposed.

The term has implications in collaborations and partnerships development from the local to the global level as well as workforce development, and national security concerns (as a shortage of STEM-educated citizens can reduce effectiveness in this area). The movement will fill the void of a less trained technical workforce and will cater to the needs of the job market. Stem psychology will enlighten learners to help raise the standard of living and will also help in getting rid of burning issues.

This book will strengthen the research in the following area:

1. Creates a universal connect among science, mathematics, and subjects of technical literacy;
2. Empowers young analytical and critical thinkers;
3. Fosters the next generation of innovators and entrepreneurs;
4. Facilitates students with opportunities to engage in 21st-century skills of teamwork, collaboration, creative thinking, and problem-solving.
5. Offers Experiential learning in which students can apply what they are learning in relevant ways.
6. Create a sense of competence, confidence, and appreciation for the application of science and mathematics in engineering and technology.

This book comprises 14 chapters. Each chapter tries to cover the various aspects of STEM –A Multi-Disciplinary Approach to achieve integrated pedagogies, innovations, and connections.

1. Empowering Future Innovators: The Importance of STEM Education in Today's World

A chapter titled 'Empowering Future Innovators: The Importance of STEM Education in Today's World' highlights the radical transformation that is currently occurring in the modern educational system. This modification impacts the teaching methods and models in addition to the curriculum, learning opportunities, assessment procedure, and syllabus. The educational system is always searching for more effective and pertinent teaching methods that take into account a variety of academic disciplines and students with different learning preferences. STEM is one of these four-discipline interdisciplinary teaching methodologies. By combining science, technology, engineering, and math to teach content, STEM education promotes an interdisciplinary approach to teaching. Through the use of STEM teaching strategies, students may be motivated to employ their problem-solving and critical thinking skills by taking part in team-based, multidisciplinary activities. Through STEM education, conventional educational procedures may be academically reorganized to incorporate twenty-first-century teaching techniques. The basics of STEM education, STEM education teaching strategies, and STEM education implementation difficulties are the key subjects of this study.

2. STEM Alignment for Industry 4.0 – A Global Overview on Education 4.0

This chapter examines and explores how technology integration into the educational field is unparalleled. Technology has influenced instruction at every level, from elementary school through postsecondary and professional education. Universities, as institutions of higher learning, have demonstrated their ongoing dedication to the development of instructional strategies and curricula that may equip the next generation of workers with the necessary information, abilities, and attitudes around the world. History attests to the fact that every previous ground-breaking industrial revolution of the modern era dictated the structure and pace of education, which transformed in line with the knowledge needed to solve real-world challenges. Industry 4.0 transformation is unavoidable and powerful, having a significant influence on institutions of higher education that are in charge of developing talent, pioneering scientific innovation, spreading knowledge, and teaching and equipping the next generation of workers. The goal of this chapter is to provide an understanding of the dimensions of Industry 4.0, Education 4.0, including STEM and its variants, and the relationships between these concepts to prepare global intellectuals with the skills and abilities necessary to succeed in the technologically disruptive world.

3. A Systematic Literature Review of STEM Education in Changing National and Global Policies and Procedures

In this chapter, important facets of STEM education are examined. STEM education is a concentrated curriculum that uses a transversal educational approach and places a lot of emphasis on science, technology, engineering, and maths. All of these topics/courses are taught to help students become more creative, critical, and logical thinkers. STEM schools and programs take an interdisciplinary approach to these significant academic fields. STEM-focused learning programs range in length from preschool to master's degree programs, depending on the resources available in a given school system or locale. STEM-based learning programs are intended to increase students' interest in pursuing further education and careers in such fields. In STEM education, a blended learning strategy is commonly employed, combining traditional classroom instruction with online learning and hands-on activities. The skills learned via STEM education expand beyond those needed to thrive in STEM fields, allowing kids with a range of interests to enter any industry and achieve by giving them practical skill sets that will enable the professional success of future generations.

4. Assessment and Evaluation of STEM Outcomes in the Education Sector

This chapter begins with a brief overview of STEM education before focusing on the emergence of common tools and related concepts in a variety of informal STEM learning environments. STEM is an interdisciplinary approach to teaching that blends theoretical concepts with real-world experiences. Students use maths, physics, and engineering in contexts that link the classroom to the outside world. The importance of STEM in education is highlighted by the Next Generation Science Standards, which were developed to improve students' understanding and application of science. Project-based learning is frequently the major emphasis of STEM education in the classroom. Use technology to emphasize the practical application of science and better prepare students for upcoming classes, projects, and activities. The chapter also highlights important advancements in education that begin with a fundamental embrace of learning objectives, such as interest, engagement, and identity construction, which go well beyond precise evaluations of knowledge and skills. It is necessary to make adjustments to the present STEM-related system to discover well-planned, organized, and interconnected policies that handle the upcoming issues, such as enhancing student outcomes, increasing the number of STEM teachers, and rethinking the curriculum.

5. Analyzing Meta Cognitive & Critical Thinking to Develop Skills and Re-Skills for STEM Literacy

This chapter is based on STEM (science, technology, engineering, and mathematics) a multidisciplinary approach that combines expertise and knowledge from several disciplines to address difficult issues and spur innovation. In a world that is changing quickly, STEM education aims to provide students with the creativity, critical thinking, and problem-solving skills they will need. Instead of discussing research in the many curriculum areas, the chapter also talks about the significance of metacognitive & critical thinking in STEM. The study focuses on how academics in the domains of science, technology, engineering, and mathematics understand the development of skills and reskills through metacognitive and critical thinking and offers succinct, pertinent research findings to enhance the caliber of integrated STEM education.

6. STEM- A Collaborative, Cooperative, and Skill-Based Learning

This chapter emphasizes the advantages of employing cooperative learning, collaborative learning, and hybrid cooperative-collaborative strategies. Working together in a small group on a structured task is cooperative learning. Each student is responsible for their work, and classes are organized into groups. When two or more individuals learn together, it is called collaborative learning. The chapter also examines the effects of skill-based learning. Instead of focusing on learning from a book, skill-based learning emphasizes teaching students practical skills that they can use in the real world. Students learning capacity is increased and their ability to function in the actual world is improved through skill-based learning. Students are challenged to think creatively and invent novel technologies as part of skill-based learning.

7. Application of Meta Cognitive, Critical Thinking, Collaborative, Cooperative, and Skill Based Learning Strategies in STEM

This chapter's primary focus is on diverse STEM learning methodologies that encourage metacognition, critical thinking, cooperation, and skill development. These many methods may be used in various educational situations and are essential components of comprehensive education. Understanding one's own thought and learning processes, as well as being aware of useful techniques for improving learning, is referred to as meta-cognitive learning. By objectively and rationally analyzing and evaluating information, critical thinking enables students to make well-informed judgments supported by logic and evidence. Working together on a common goal while promoting communication, and the sharing of information and skills is the essence of collaborative learning in the STEM fields. Similar to collaborative learning, cooperative learning emphasizes individual accountability and the need to contribute to the achievement of the group as a whole. Focus is placed on developing certain characteristics including leadership, communication, and problem-solving abilities. Students' understanding and memory of the material can be improved by combining these various learning approaches into STEM education. The study will also address developing their problem-solving and decision-making skills, which will help them succeed in the job market of the twenty-first century.

8. Exploring the Effectiveness of Coding Education Methods and their Impact on the Data Science Job Market Requirements – A Systematic Review

The idea of data science and coding is the highlight in this chapter which connotes that number of school programs in data science and coding has increased as a result of the rising demand for these talents. The objective of this systematic review is to assess the body of knowledge on the influence of coding education on the advancement of the data science profession. To find and evaluate pertinent studies on data science and coding education, the study will adopt a strict and methodical methodology. Using keywords and MeSH phrases associated with data science, coding education, career development, and program assessment, a thorough search of pertinent databases will be carried out. This systematic review's objective is to assess the body of knowledge on the usefulness of coding instruction for the advancement of data science careers. The number of school programs in these fields has increased as a result of the rising need for people with data science and coding expertise.

9. Utilization of Science Parks, Museums, Zoos, and Aquariums (SPMZAs) for Enriching STEM Education in Developing South Asian Countries in Today's Digital World

The term STEM emphasizes that studying the four subjects of Science, Technology, Engineering, and Mathematics essential for the advancement of both technology and the economy. As a result, India has a relatively large number of higher education institutions that provide a wide range of STEM courses. The enormous number of students who choose to enroll in these courses year after year is proof of their popularity. Lack of resources (financial, technological, and human) is the major obstacle to offering these courses in emerging South Asian nations. Collaboration between educational institutions and public or private science parks, museums, zoos, and aquariums (SPMZAs) can significantly lessen this. To facilitate and improve STEM education, this chapter considers how educational institutions might work with the SPMZA facilities already in place in various regions of South Asia. This study is a fantastic contribution to the idea of equal opportunity in schooling. We also explain how the use of virtual reality and augmented reality has improved our hypotheses, and we propose an adaptable model based on them.

10. Measuring the Effectiveness of Deep Learning in STEM Education: A Comparative Analysis of Student Outcomes and Engagement

The chapter investigates how Deep learning algorithms have drawn more attention in recent years as a way to improve STEM teaching. By comparing student results and engagement, The study analyses the effectiveness of deep learning in STEM education. By comparing student results and engagement, this study seeks to determine the efficacy of deep learning in STEM education. A kind of machine learning called deep learning has demonstrated significant promise in many areas, and recent years, its use in education has drawn attention. The study also investigates how student learning results and engagement in STEM courses are impacted by deep learning methodologies. Student results are assessed using a variety of assessment instruments, such as examinations, projects, and performance reviews, with the use of deep learning. Student motivation, involvement, and interest in the learning process are all measured through questionnaires and observations to determine engagement. The group that engaged in deep learning showed greater levels of enthusiasm, active participation, and interest in STEM fields.

11. Employability Challenges and Opportunities under STEM Learning with Special Focus on Apprenticeship & Internship

This chapter states that employability is a major problem. Very few candidates have the mindset required to quickly pick up the requisite skills and knowledge, and the skills and knowledge they do have are frequently not applicable to the sector. STEM education is practically required for future employment. It is also true that different jobs and activities require various levels of STEM application knowledge, ability, and experience. Different degrees of employment and different technologies within the same business lead to varying demands for STEM application within the workforce. Internships and apprenticeships can assist people and certain sectors in designing and delivering their special STEM needs among potential employees in this situation. While the general job market needs are taken into consideration while designing formal education curricula, internships can provide the unique KSE (Knowledge, Skill, and Experience) requirements.

12. Moving Towards Practicality in Higher Education through STEM Education and Work-Based Learning

This chapter explores that by escaping the confines of traditionalism, following the paths of innovation, and developing complex contextual solutions using the tools and technology of today, STEM fields have evolved into a meta-discipline. It has entered the race to gain economic significance and become internationally recognized in the global economy of the twenty-first century. Teachers from throughout the community may collaborate to develop effective pedagogical models that enhance the comprehension of the STEM curriculum. The suggested study intends to comprehend and spread the benefits of work-based learning in general and STEM in particular. Additionally, it aims to outline the fundamental characteristics of STEM and the future development of the field in third-world countries like India.

13. Frugal Innovation: Lessons from Successful Entrepreneurs and Trends Shaping STEM Future

This chapter looks that in recent years, particularly in emerging economies, the idea of frugal innovation has become more and more popular. It entails developing novel goods, services, or business models while working with constrained resources, and many entrepreneurs have utilized it to build successful enterprises. This chapter also looks at some case studies of prosperous businesspeople who have employed frugal innovation tactics, analyzing their strategies and difficulties, and emphasizing insightful information for businesspeople wishing to adopt comparable methods. Additionally, it will talk about how government policies and education can support the growth of entrepreneurship and cost-effective innovation.

14. STEM Employees: Role of Co-Worker Support, Perceived Insider Status and Organisation Citizenship Behaviour in Achieving Employees' Retention

This chapter is based on a survey of 300 STEM workers in Madhya Pradesh, India. The study focuses on how organisational civic conduct, perceived insider position, and coworker support play a role in STEM employee retention. Organisations must make an effort to concentrate on characteristics like organisational citizenship behaviours, perceived insider status, and coworker support, as well as on other elements like incentives and recognition. To keep the employees in the company, they should make an effort to create a pleasant work atmosphere. The results show a positive association between coworker support and organisational citizenship behaviour as well as a favourable relationship between perceived insider status and OCB and employee retention in the company.

Priya Jindal
Chitkara Business School, Chitkara University, Punjab, India

Amandeep Singh
Chitkara Business School, Chitkara University, Punjab, India

Rashmi Aggarwal
Chitkara Business School, Chitkara University, Punjab, India

Chapter 1

Empowering Future Innovators: The Importance of STEM Education in Today's World

Renjith V. Ravi[1,*]**, PhD**
Govind Shinde[2]**, PhD**
T. V. Roshini[3]**, PhD**
and Saw Mon Yee Aung[4]**, PhD**

[1]Department of Electronics and Communication Engineering, M.E.A Engineering College, Malappuram, Kerala, India
[2] H.K. Institute of Management Studies and Research, Mumbai, Maharashtra, India
[3]Vimal Jyothi Engineering College, Kannur, Kerala, India
[4]Department of Information Technology Engineering, Technological University (Toungoo), Toungoo, Myanmar

Abstract

The contemporary educational system is going through a period of radical change. This change affects not only the curriculum, learning opportunities, assessment process, and syllabus but also the teaching models and techniques. The educational system is always looking for more efficient and relevant teaching strategies that incorporate several academic fields and students with various learning preferences. One of these teaching strategies that blends four disciplines is STEM. STEM education encourages an interdisciplinary approach to teaching by fusing the fields of science, technology, engineering, and math to impart subject matter. Students may be inspired to use their problem-solving and critical

---

[*] Corresponding Author's Email: renjithravi.research@gmail.com.

In: STEM
Editors: Priya Jindal, Amandeep Singh and Rashmi Aggarwal
ISBN: 979-8-89113-305-1

thinking abilities by participating in team-based, interdisciplinary assignments through the use of STEM teaching techniques. Traditional educational approaches may be academically restructured to include the twenty-first century teaching methods via STEM education. STEM education uses a variety of teaching techniques, including problem-based learning, inquiry-based learning, project-based learning, cooperative learning, and more. The main topics of this study include the fundamentals of STEM education, STEM education teaching methods, and STEM education implementation challenges.

Keywords: STEM education, teaching models and techniques, interdisciplinary approach

Introduction

With the advent of globalization, technological advancement, and computerization scraping several conventional employment terminologies, there is an imminent need to reconsider the direction of our education system and implement it as soon as possible if we would like our nation to continue to play a significant role in the quickly changing global realities (Li et al., 2020).

Despite possessing exceptional potential, these students' ability to innovate, solve problems, and be creative has been constrained by the previous educational approach, which was exam-focused. A paradigm shift in education is required to educate students to be globally aware while giving them the skills and talents they need to thrive and survive in the new social and economic order since the whole globe is nearly becoming a linked global village. In order to meet the demands of the future industry, a well-designed STEM (science, technology, engineering, and mathematics) curriculum should be developed and used in schools. Huge prospects are approaching our nation's doors as a result of globalization (Chesky and Wolfmeyer, 2015). We can definitely take advantage of the opportunity provided by the current situation since we are the second-most populous nation in the world, possess unparalleled talent and culture, and have the biggest youthful population anywhere on Earth. To take advantage of the chances and advantages that STEM education can provide our nation, India requires the cooperation of both the government and other educational groups. STEM employment is increasing at its quickest pace ever. The National Research Council (NRC, 2014) predicts that 80% of the occupations that are produced in the next ten years will require some level of math and scientific expertise. It is permissible to refer to these abilities as 21st-century or STEM abilities. Now that the Indian government is concentrating on initiatives like the "Make in India" Innovation Mission, innovation and manufacturing are being developed in schools. The time is now for India to accept the challenge and foster an innovation- and application-based learning culture among its institutions, universities, and instructors. Within the Indian education sector, the number of STEM jobs is growing far faster than the number of STEM graduates. According to research, kids start showing interest in STEM areas around the age of eight (Gao et al., 2020).

This is a result of their fascination with technology, gadgets, and gizmos. But when it does happen, it is a very gradual change from being a technology consumer to an inventor. Evidently, India lacks the connection between engineering, technology, and entrepreneurship. Since a few years ago, certain institutions that were early adopters of STEM have been adopting cutting-edge methods to make sure that kids are more disposed to and fascinated by the "do it yourself" way (Amirtham and Kumar, 2021; Krishnan and Hariharan, 2016; Sarukkai, 2015). Before India can fully take advantage of these international prospects, there are a few really important obstacles that must be solved in order to scale up the development of STEM talents. Making sure that youngsters get the greatest possible supervision and assistance is one of the main problems associated with implementing STEM education.

Another major obstacle is our school system's unwavering emphasis on grades, which is exam-focused and exam-oriented. Our methods for teaching, learning, and evaluating need to be drastically changed. Many instructors continue to believe that adding STEM would distract pupils from their studies and prevent them from finishing their assigned curricula in the allotted time. The only way to deal with this is to make them aware of it and persuade them otherwise by demonstrating the advantages of exposing their children to STEM learning approaches. Through the introduction of national curricula that provide frameworks and projects that use the STEM approach, nations all over the globe are embracing this technique. The quality of the infrastructure and the money required to build it represent the third major obstacle to STEM education (Li et al., 2019). Here, the business community must work together and provide support for school improvement via their CSR initiatives. Over the years, the cost of doing high-quality research and education has skyrocketed, and cutting-edge technologies need cutting-edge facilities. Several nations have convinced their local governments, regions, and institutions to make the switch after understanding the possibilities of a STEM curriculum (Lee et al., 2019). There will probably be a need for tech expertise across industries due to developments in a number of scientific and technological fields, including robots, the Internet of Things, and so on. To be competitive in the market, job seekers must constantly upgrade their abilities since the prerequisites for these positions get more complex with time. With the aid of Atal Tinkering Labs (ATL), the Indian government has helped more than 3000 schools' outfit themselves and enable students to get practical experience with cutting-edge technology as part of the "Skill India" initiative. Robotics is a key component of STEM education because it offers an intriguing way to incorporate STEM subjects into the classroom. It is past time we abandoned our antiquated educational system and gave kids the abilities they need to thrive in the twenty-first century, including critical thinking, creativity, and invention.

The minimum standard of education for every student would then be to have a strong foundation in math and science as well as imaginative thinking, interpersonal interaction, and teamwork abilities. The ideal way to accomplish these goals in a methodical way is via STEM (science, technology, engineering, and mathematics).

The requirement for qualified instructors who are knowledgeable about the most recent advancements in their professions is also emphasized by STEM education (Li et al., 2019). This is yet another significant obstacle to the successful execution of STEM-based education in our nation. Because of how big the issue is, it cannot be resolved without a concerted effort by the government, businesses, society, and the general populace. We barely have a choice but to avoid this issue if we wish to have a substantial impact on the global order in the next few decades, despite its magnitude and complexity. Let's work together to advance the cause of worthwhile education, which has the potential to bring wealth and well-being to both our next generation and the global community as a whole.

What Exactly Is Meant by STEM?

STEM stands for science, technology, engineering, and mathematics subjects in education. Classes provide a strong focus on real-world applications and give students the practical knowledge they need to succeed in the STEM field (Margot and Kettler, 2019).

Learners should be capable of applying a broad range of transferrable abilities, such as entrepreneurial endeavours and problem-solving skills, to their employment prospects at the conclusion of their study. The idea of STEM is depicted in Figure 1 and the depiction of the concept as a tree with leaves and branches is shown in Figure 2.

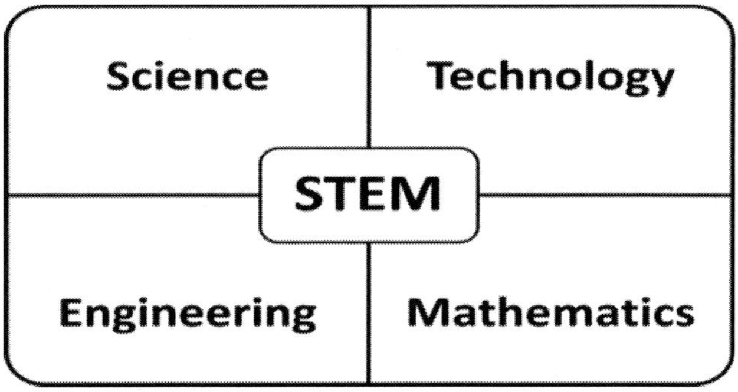

Source: Author's compilation.

Figure 1. STEM courses.

Science

The natural and physical worlds are studied through observation and experimentation in a science-based curriculum. Scientists often do research, write scholarly articles and proposals, and present their results (Nugroho et al., 2019). On the other side, scientific specialists often help scientists by carrying out tests, collecting samples, and more. By the specific area of research, scientific areas are classified, such as:

- Physical Sciences
- Chemistry
- Biological Sciences
- Earth science
- Social Sciences

Technology

Technology-based curricula integrate engineering and scientific research to build systems for information and computing (Manosuttirit, 2019; Pleasants et al., 2019). Technical education often entails developing, evaluating, advancing, and maintaining software, hardware, systems, and networks. It also allows communication. Artificial intelligence, computers, operating systems, autonomous systems, computer science, secure communication, and computer programming are examples of fields that heavily rely on technology.

Engineering

Engineers discover answers to real-world issues by fusing technology, math, and science. Typically, engineering work involves the creation of goods, materials, systems, and frameworks (Wang and Chiang, 2020; Hasanah, 2020). The industries, in which they are used, such as oil and gas, textiles and clothing, or aviation, tend to classify engineering specialties, which include:

- Civil Engineering
- Mechanical Engineering
- Electrical Engineering
- Electronics Engineering

Mathematics

Applications in the mathematics faculty analyze and resolve issues using logical, spatial, and numerical linkages (Maass et al., 2019). The other three STEM areas have a technical basis in math. It often entails finding patterns in data or abstract logic. Students may model issues, examine linkages, and draw conclusions using these patterns. Typical math specialties include:

- Algebra
- Geometry
- Statistics
- Game theory
- Trigonometry

STEM Disciplinary Integration

There are several ways to interpret the term "integration". For instance, Hurley (2001) carried out a meta-analysis and presented mathematics and scientific success metrics while comparing integrated mathematics and science education to a non-integrated control group (Srikoom and Faikhamta, 2018). Hurley used the following criteria to divide the accomplishment scores by integration level:

- *Sequential:* Science and mathematics are taught and prepared in a sequential order, one coming before the other.
- *Parallel*: Using parallel principles, science, and mathematics will be organized and taught concurrently.
- *Partial*: In the same classrooms, science and mathematics are provided with instruction in both combined and separate settings.
- *Enhanced*: The core subject of education is either science or mathematics, with hints of the other subject throughout the lesson.

Source: Author's compilation.

Figure 2. The concept of STEM as a tree.

Overall, science and math are taught side by side with the goal of equality.

Additionally, according to National Research Council (2014), the word "integrated" is used indiscriminately in educational practices and research and is often used interchangeably with phrases like "connected," "unified," "interdisciplinary," "multidisciplinary," "cross-disciplinary," and "transdisciplinary." Because links might appear at several levels at once, expressing integrated STEM education is made more challenging.

The idea of degrees of STEM discipline integration was previously applied to professional development courses and utilized as illustrations of integrated STEM education in the classroom for participants to use as a primary guideline when developing their own realistic STEM experiments or instructional plans. Therefore, it is possible to combine and categorize Honey's degree of integration and Hurley's level of integration into the following groups as shown in Figure 3.

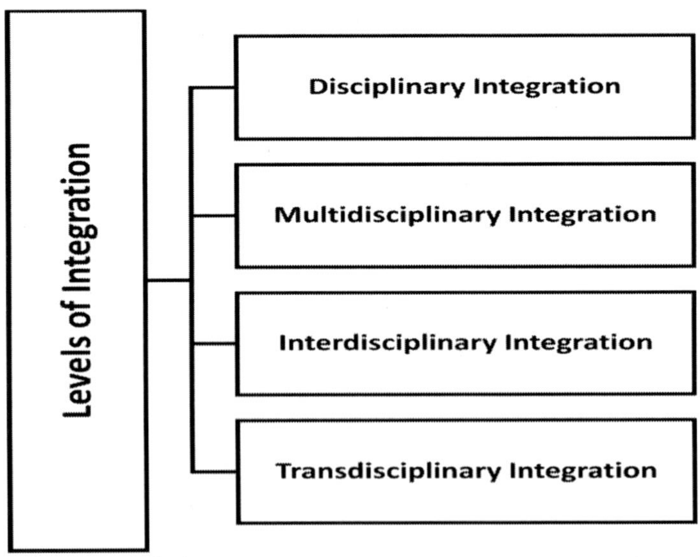

Source: Author's compilation.

Figure 3. Levels of integration in STEM.

- *Disciplinary Integration*: The study of science, technology, engineering, and mathematics is organized and taught independently in each field.
- *Multidisciplinary Integration*: Multidisciplinary integration is the planning and teaching of science, technology, engineering, and math

disciplines sequentially and individually while maintaining a common theme.

- *Interdisciplinary Integration*: Interdisciplinary integration refers to the explicit connections between the scientific, technological, engineering, and math disciplines, which are planned and taught independently via education.
- *Transdisciplinary Integration*: Science, technology, engineering, and math disciplines are organized and taught harmoniously in the same environment via transdisciplinary integration.

Perspectives STEM Integration

Furthermore, Bybee (2013) said that there are several viewpoints on STEM integration at the integration level. For instance, in STEM Equal, a Quartet of Separate Disciplines shown in Figure 4 (a), and STEM as a Reference for Science and Mathematics depicted in Figure 4 (b), both viewpoints included the same four components, but they differed in how they interacted with one another. First, STEM is seen as a distinct idea with no clear connections. This might be a single course that covers all of the STEM fields, or it can be four different courses—one for each subject. The second, STEM is seen as a course that focuses only on science and mathematics, with or without any mention of engineering and technology.

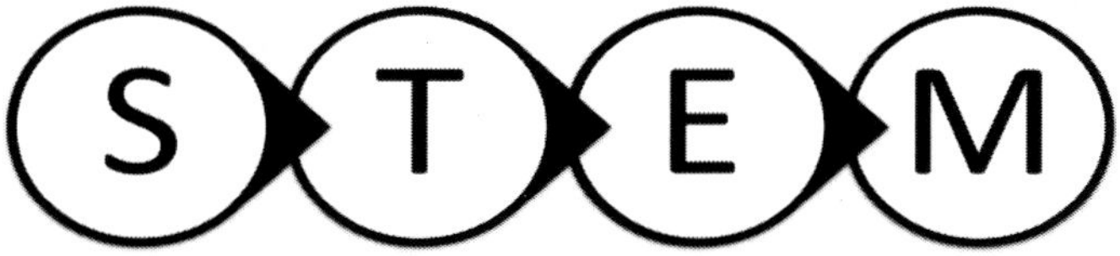

Source: Dheesha and Parween (2022).

Figure 4 (a). STEM as an equal quartet of individual.

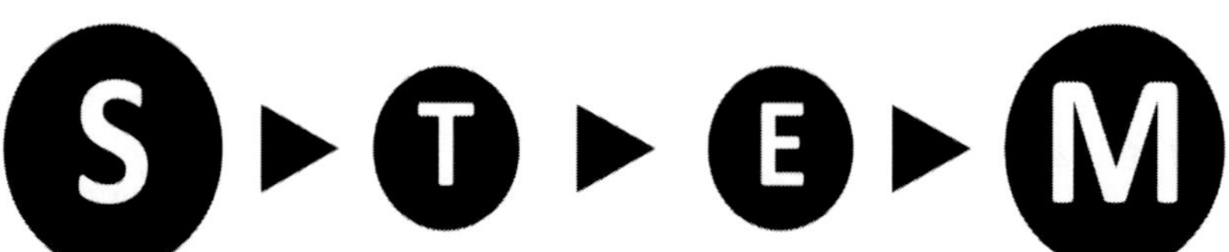

Source: Dheesha and Parween (2022).

Figure 4 (b). STEM with high priority for science and mathematics.

The Importance of STEM

Students who study STEM in higher education learn how to think critically, are better equipped for the workforce, and become experts who can collaborate across scientific fields to tackle complex challenges (Ryoo and Calabrese, 2018). STEM combines a number of academic fields and teaches students how to solve issues by combining their disciplinary expertise. STEM students also take challenging and diverse courses to develop a strong work ethic.

STEM education encourages a do-as-you-learn philosophy. Students take part in initiatives that have real-world repercussions. For instance, students often take part in academic research initiatives. With this strategy, students may study while actively participating in the advancement of cutting-edge research and new technologies.

STEM-focused universities also gain from hosting these initiatives. The innovative technology that the programs produce earns them money and praise. For instance, the National Institute of Allergy and Infectious Diseases awarded the University of Minnesota $66 million to support a project for the development of antiviral drugs in 2022. This scholarship was given to the institution in recognition of its creative responses to the COVID-19 epidemic.

The Functioning of STEM

The main goal of STEM disciplines and curricula is the development of transferrable 21st-century skills. Students have the opportunity to develop their soft skills, such as teamwork and communication, throughout middle and high school (Kaleci and Korkmaz, 2018). Large-scale projects and examinations are used to develop and put these abilities into practice in post-secondary education. College degrees place a strong emphasis on educating students for STEM careers in the real world and advancing their professional growth in this field.

The four Cs of 21st-century education creativity, collaboration, analytical thinking, and communication are embraced by STEM learning. For the complicated problems that workers in STEM-related sectors will unavoidably encounter, these four domains are crucial. Students gain the ability to approach problem-solving from many perspectives by using critical thinking and creative abilities. On the other hand, students learn how to spot mistakes, gather relevant data, and properly understand and apply it to different systems.

Students who study STEM also learn how to create and find cost-effective solutions using novel or previously unconsidered methods.

Students can easily perceive VSEPR models and challenging chemical compounds thanks to Active Chemistry. California State University, Long Beach educator Elaine Bernal utilizes the platform to help students understand the significance of their education on a larger scale. The platform encourages students to learn more about chemistry by keeping them interested visually. Genuine environmental challenges, such as global warming, water chemistry, and climate change. When kids observe how it works, they become more aware of it, according to her. In STEM fields, communication takes place both orally and in writing. For ideas to be successfully translated into action, this involves technical writing, interpersonal interaction, addressing audiences, and even teaching.

Internships, research, and volunteer activities are often included in or made available by STEM curricula in order to further professional growth. Additionally, internships assist students in applying their newly gained knowledge and abilities to their chosen sector. Here are some benefits of doing a STEM internship now for college students:

- Students enhance their resumes and gain highly desirable abilities.
- The opportunity to experience a full-time profession in their field of interest is provided to students.
- Students get the opportunity to network and make new contacts for their future professional lives.

Competitions and Games

While online environments may surely accommodate games and contests, there is much to be learned from the tactile sensation of competing against others in one's close vicinity (Wang et al., 2022; Donmez et al., 2020; Gao et al., 2020). In-person settings may be used to help students more completely explore, absorb, and apply complicated information because of the group dynamic's sheer spontaneity and collaborative energy, which can be challenging to mimic online.

STEM and Online Education

When STEM education became totally online in 2020, teachers and students were confronted with new difficulties. Using technology alone has made distractions easier (Chirikov et al., 2020). Digital learning has also made it more challenging to communicate with classmates and lecturers. Many STEM courses use lab-based learning, which presents substantial engagement issues (Kefalis and Drigas, 2019). Some colleges have enhanced their laptop lending services and Wi-Fi hotspot access. To meet the needs of the distant learning environment, teachers have also created innovative teaching strategies and activities (Prince et al., 2020).

Lab Experiments at Home

Teachers may give lab kits to kids for non-hazardous activities and supplies so they can do experiments at home. Students may use a webcam to show their comprehension of a subject or use a live video stream to follow along with the instructor (Rao and Dave, 2019). This strategy is one more tool to support fairness and access in education. Students may still engage in hands-on learning without physically traveling to school by having materials sent to them.

Data Analysis and Research

Research and data analysis projects are perfect for online learning, even if many activities may not be suitable (Zhao, 2022; Nikalus et al., 2019). Give your pupils a dataset and instruct them to analyze it using a set of guidelines or reasoning. To help pupils improve their writing abilities, you may also have them write a reflection paper as a follow-up.

Lectures by Outside Speakers

One benefit of online instruction is that participants from all around the globe may attend your lesson. It also becomes simpler for you or your teaching assistants to plan and allocate a budget (Belser et al., 2018). Similar to live webinars, STEM learning may be improved by making it more interactive and interesting. Consider awarding each student who participates in a webinar one or two additional points.

STEM and Blended Learning

In recent years, it has become clear that in-person and online learning are the most effective. As technology adapts and develops, the blended classroom, where instruction is distributed evenly across in-person and online locations, has become the standard (Allan et al., 2019). No matter where classes are held, instructors must establish effective teaching strategies to keep the learning process new and interesting.

Including components of digital learning, like research work, in your in-person seminars and practical projects is a useful approach to doing this (Capone, 2022; Owston et al., 2020). The encouragement of experimentation encourages pupils to investigate fresh and original solutions to issues. In-person and online learning may be combined via cooperative activities like reading the laboratory manual in pairs and responding to pre-lab questions.

STEM-Based Teaching Technique

Diverse topic knowledge and skills are included in STEM education in order to apply concepts to an idea, a product, or an operation. Science, technology, engineering, and math are often referred to as STEM. Here, concerns concerning the integration of all four disciplines into carrying out the specific action emerge. Equal integration of all four disciplines is not always attainable. The practice of teaching may entail three of the four disciplines.

Only a few different teaching approaches may be employed to teach material since STEM is a distinctive idea. The project-based teaching approach, the discovery approach, the cooperative approach, the inquiry-based approach, and the problem-based approach are all sometimes used. Based on the aforementioned principles, the problem-based teaching approach is a popular STEM teaching strategy. The efficiency of the aforementioned teaching techniques as a means of learning for students of various age groups and levels may be determined by a variety of study methods. Nguyen and Bui

(2023) found that the problem-based learning approach significantly increased students' learning results and effective teaching strategies.

Inquiry-Based Teaching Method

This approach to education is built on challenging any social or environmentally related issue, evaluating potential answers critically, and addressing issues by putting ideas into practice. Teachers must cover topics such as catastrophe preparedness, biotechnology, economics, and energy. The following stage of this methodology is followed by a round of questions and answers. In this situation, the instructor serves as a scaffold for the students, offering them help and support while helping them find solutions to the difficulties posed. The context-based viewpoint is used in the teaching of STEM subjects. The teaching strategy consists of the following seven stages (Sheth and Pathak, 2023; Sutaphan and Yuenyong, 2019):

1) Identifying societal problems,
2) identifying a feasible remedy,
3) knowledge need, and
4) decision-making
5) Creation of a product or prototype;
6) Testing and assessment of the solution
7) The socialization and decision-making stages

Project Based Method

This form of education is used in the teaching of STEM subjects. This approach involves the instructor giving the class a project to work on and the students using what they have learned in class to accomplish it. Students work on projects that are based on real-world scenarios. The teacher's job will be to serve as a facilitator and motivator, offering instruction and encouragement to the pupils. Even though this approach is more expensive and calls for more supplies and materials, it works very well when teaching young STEM students. With this approach, kids' collaborative skills, sensory, cognitive, and critical thinking capabilities all increase.

Cooperative Teaching Method

When students must collaborate in groups to finish a particular job, the cooperative teaching method—which is not always used as a separate teaching strategy—indirectly merges with the other three. Many students collaborate and support one another to finish a task in this participatory learning technique. According to Hanh (2021), students grow in their ability through cooperation, have just encouraged the ability to work together among team members, and have just promote a beneficial influence on each other. In (Johnson and Johnson 1998) (Johnson et al., 1998), the specifics of the instructor's involvement in setting up formal cooperative learning groups are described. (1) state the lesson's objectives; (2) make a number of instructional decisions (such as group size and how students are divided into groups); (3) describe the task and its positive interdependence; (4) keep an eye on students' learning and get involved in the groups to help with the task or improve teamwork skills; (5) assess students' learning and assist students in reflecting on how well their group performed (Romer, 2000). In STEM classroom training, this thorough instructional methodology is frequently employed.

Problem-Based Method

Higher-order thinking is the foundation of this instructional approach. Students must once again put forth the effort to solve the teacher's dilemma. Students must put established ideas and concepts from multidisciplinary research that span several scientific domains to use. To solve an issue, they must collaborate in groups, use their knowledge, and use critical thinking and situational analysis.

The problem-based approach is sometimes confused with the inquiry-based approach. The important distinction between the two approaches is that the inquiry-based approach requires more individualized student effort. In contrast, students may collaborate in groups while using the problem-based approach. The problem-based approach poses several open-ended challenges that may not have a clear answer. Students must come up with interim fixes or take safeguards for problem management for such issues.

Significance of STEM Education

STEM education offers a radically novel method of instruction that enables students to apply their understanding of four academic fields (science, technology, engineering, and mathematics) and lengthens and enriches the learning experience. STEM has a good link with 21st-century abilities including communication, teamwork, critical thinking, and creativity, sometimes known as the "4C skills," according to a study by (Diana and Turmudi, 2021).

There is a clear need to change educational strategies in order to teach the next generation how to think creatively and innovatively and how to solve problems by considering many options. Adopting contemporary strategies like STEM (science, technology, engineering, and mathematics) education may make this achievable. STEM education is crucial since it promotes the advancement of science and technology as well as sustainable growth, according to (Kaleci and Korkmaz, 2018).

The world is moving towards automation with the introduction of cutting-edge technology like artificial intelligence, robots, and the Internet of Things. The lack of STEM expertise in India may surprise many people given its rich heritage. While many of the best talents on the globe come from India and are well-known for their IT abilities, the demand for such "STEM-skilled" labour in our own nation considerably exceeds the supply (Mishra, 2022). Their critical thinking and problem-solving abilities are developed through STEM education, which also encourages them to think creatively. They may work in a variety of fields and use their cross-disciplinary expertise to further their careers. The economy may benefit from these workplace advances.

STEM is required to satisfy the needs of workplaces in the twenty-first century due to the fast expansion of ICT and globalization, which are reshaping society (Salic-Hairulla and Monera, 2021). The National Science Foundation (NSE, 2018) forecasts that 80% of the occupations that are produced in the next ten years will require some level of math and scientific expertise. At the secondary level, the curriculum has to be modified to raise students' scientific literacy (Mishra, 2022). STEM Connector research from 2018 estimates that 2.4 million people are needed globally to fill STEM-related professions (Dheesha and Parween, 2022). The studies mentioned above highlight the wide range of STEM education across nations, with plenty of career opportunities for students in the next few years.

Numerous scholars have discovered STEM instruction and STEM teaching methods in other nations. But there wasn't much research done in

India. In India, there aren't many schools that prepare instructors to teach STEM subjects. It is still in its early stages and at an experimental level, however. More research is needed in this area, including how to change the curriculum to incorporate STEM education, how to evaluate STEM education, how to create textbooks for this purpose, and how to foster positive mindsets and consciousness.

Conclusion

Implementing STEM (Science, Technology, Engineering, and Mathematics) is one of the teaching strategies that is seen as suitable for preparing students' knowledge and abilities for the rapidly expanding industrial age. Science, technology, engineering, and math (STEM) education is an initiative to include some or all of these four disciplines into a single class, unit, or lesson that is centered on links between the topics and real-world issues. STEM education seeks to improve critical thinking abilities in order to better prepare students to become innovators in a constantly changing environment and to solve this problem in subsequent educational endeavors by including them in the study of science, technology, engineering, and math.

References

Allan, C. N., Campbell, C., & Crough, J. (2019). Blended learning designs in STEM higher education. *Blended learning designs in STEM Higher Education*, 339-362.

Amirtham S, N., & Kumar, A. (2021). Gender parity in STEM higher education in India: A trend analysis. *International Journal of Science Education*, *43*(12), 1950-1964.

Belser, C. T., Prescod, D. J., Daire, A. P., Cushey, K. F., Karaki, R., Young, C. Y., & Dagley, M. A. (2018). The Role of Faculty Guest Speakers and Research Lab Visits in STEM Major Selection: A Qualitative Inquiry. *Journal of Career and Technical Education*, *33*(1), 8-26.

Bybee, R. W. (2013). *The case for STEM education: Challenges and opportunities*. USA, National Science Teachers Association Press.

Capone, R. (2022). Blended learning and student-centered active learning environment: a case study with STEM undergraduate students. *Canadian Journal of Science, Mathematics and Technology Education*, *22*(1), 210-236.

Chesky, N. Z., & Wolfmeyer, M. R. (2015). *Philosophy of STEM education: A critical investigation*. Palgrave Macmllian USA, Springer.

Chirikov, I., Semenova, T., Maloshonok, N., Bettinger, E., & Kizilcec, R. F. (2020). Online education platforms scale college STEM instruction with equivalent learning outcomes at lower cost. *Science advances*, *6*(15), 5324.

Dheesha, J. B., & Parween, S. (2021). Adapted Science Instructional Aids–An Anchor For Children With Visual Impairment. *Journal of Disability Management and Special Education*, *4*(2), 10.

Diana, N., & Turmudi, T. (2021). Teachers' Readiness in Developing STEM-based Module to Support Teaching and Learning in the 21st Century. *Edumatica: Jurnal Pendidikan Matematika*, *11*(2), 1-8.

Dönmez, İ., Tekce, M., & Kirmit, Ş. (2020). Using digital games in technology oriented STEM education: The examination of the students' game designs. *Journal of Education in Science Environment and Health*, *6*(2), 77-91.

Gao, F., Li, L., & Sun, Y. (2020). A systematic review of mobile game-based learning in STEM education. *Educational Technology Research and Development*, *68*, 1791-1827.

Gao, X., Li, P., Shen, J., & Sun, H. (2020). Reviewing assessment of student learning in interdisciplinary STEM education. *International Journal of STEM Education*, *7*(1), 1-14.

Hanh, P. T. H. (2021). Develop cooperative capacity for students in STEM modelation model. In *Journal of Physics: Conference Series*, 1835(1), 012054, IOP Publishing.

Hasanah, U. (2020). The effectiveness of STEM education for overcoming students' misconceptions in high school physics: Engineering viewpoint. *Science Education International*, *31*(1), 5-13.

Hurley, M. M. (2001). Reviewing integrated science and mathematics: The search for evidence and definitions from new perspectives. *School science and mathematics*, *101*(5), 259-268.

Johnson, D. W., Johnson, R., & Smith, K. (1998). Active learning: Cooperation in the college classroom . Edina, MN: Interaction Book Company.

Johnson, D. W., Johnson, R. T., & Smith, K. A. (1998). Cooperative learning returns to college what evidence is there that it works?. *Change: the magazine of higher learning*, *30*(4), 26-35.

Jindal, P., & Tej, F. (2016). Being an educator in private vs. public school system: A comparative study on job contentment. *IRA International Journal of Education and Multidisciplinary Studies*, *3*(3), 309-319.

Kumari, S., Jindal, P., & Chand, P. K. (2022). Social Media Networking: As Facilitator of Acculturation. *ECS Transactions*, *107*(1), 9381.

Kaleci, D., & Korkmaz, Ö. (2018). STEM Education Research: Content Analysis. *Universal Journal of Educational Research*, *6*(11), 2404-2412.

Kefalis, C., & Drigas, A. (2019). Web Based and Online Applications in STEM Education. *Int. J. Eng. Pedagog.*, *9*(4), 76-85.

Krishnan, P., & Hariharan, S. (2016). *Challenges in STEM education for'Skill India'.* Nature Publishing Group.

Lee, M. H., Chai, C. S., & Hong, H. Y. (2019). STEM education in Asia Pacific: Challenges and development. *The Asia-Pacific Education Researcher*, *28*, 1-4.

Li, Y., Schoenfeld, A. H., diSessa, A. A., Graesser, A. C., Benson, L. C., English, L. D., & Duschl, R. A. (2019). Design and design thinking in STEM education. *Journal for STEM Education Research*, *2*, 93-104.

Li, Y., Wang, K., Xiao, Y., & Froyd, J. E. (2020). Research and trends in STEM education: A systematic review of journal publications. *International Journal of STEM Education*, *7*(1), 1-16.

Maass, K., Geiger, V., Ariza, M. R., & Goos, M. (2019). The role of mathematics in interdisciplinary STEM education. *Zdm*, *51*, 869-884.

Manosuttirit, A. (2019). How to Apply Technology in STEM Education Activities. In *Journal of Physics: Conference Series*, 1340 (1), 012007. IOP Publishing.

Margot, K. C., & Kettler, T. (2019). Teachers' perception of STEM integration and education: a systematic literature review. *International Journal of STEM education*, *6*(1), 1-16.

Mishra, P. (2022). Scientific Literacy: To days Necessity: A Comparative Study of 10th Class Students from Different Schools. *International Journal of Humanities and Education Development (IJHED)*, *4*(1), 39-42.

Nguyen, N. D., & Bui, L. T. H. (2023), Enhancing STEM teaching competency for mathematics teachers in Vietnam. In *AIP Conference Proceedings*, 2685 (1). AIP Publishing.

Nikalus, S. S. L., Kee, S. E., Chau, G. H., & Ooi, P. L. (2019). Data Analytics for Engagement In Stem Education. In *Proceedings of the 4th International Conference on Big Data and Computing*, 60-63.

National Research Council. (2014). *STEM integration in K-12 education: Status, prospects, and an agenda for research*. National Academies Press.

Nugroho, O. F., Permanasari, A., & Firman, H. (2019). The movement of stem education in Indonesia: Science teacher's perspectives. *Jurnal Pendidikan IPA Indonesia*, *8*(3), 417-425.

Owston, R., York, D. N., Malhotra, T., & Sitthiworachart, J. (2020). Blended Learning in STEM and Non-STEM Courses: How Do Student Performance and Perceptions Compare?. *Online Learning*, *24*(3), 203-221.

Pleasants, J., Clough, M. P., Olson, J. K., & Miller, G. (2019). Fundamental issues regarding the nature of technology: implications for STEM education. *Science & Education*, *28*, 561-597.

Prince, M., Felder, R., & Brent, R. (2020). Active student engagement in online STEM classes: Approaches and recommendations. *Advances in Engineering Education*, *8*(4), 1-25.

Ryoo, J. J., & Calabrese Barton, A. (2018). Equity in STEM-rich making: Pedagogies and designs. *Equity & Excellence in Education*, *51*(1), 3-6.

Romer, P. M. (2000). Should the government subsidize supply or demand in the market for scientists and engineers?. *Innovation policy and the economy*, *1*, 221-252.

Rao, A. R., & Dave, R. (2019). Developing hands-on laboratory exercises for teaching STEM students the internet-of-things, cloud computing and blockchain applications. In *2019 IEEE Integrated STEM Education Conference (ISEC)*, 191-198. IEEE.

Salic-Hairulla, M. A. (2021). Utilization of ICT-integrated STEM lessons on biodiversity for grade 8 learners. In *Journal of Physics: Conference Series,* 1835(1), 012074. IOP Publishing.

Sheth, M. M., & Pathak, R. (2023), Stem Education: An Interdisciplinary And Integrated Approach Of Teaching. *Interdisciplinary Approaches and Strategies for Sustainable Development*, 1, 80-87.

Sarukkai, S. (2015). Challenges for STEM education in India. In *International Science and Technology Education*, 95-110. Routledge.

Srikoom, W., & Faikhamta, C. (2018). Assessing in-service teachers' self-efficacy and beliefs about STEM education. *Journal of Education, 12*(4), 170-186.

Sutaphan, S., & Yuenyong, C. (2019, October). STEM education teaching approach: Inquiry from the context based. In *Journal of Physics: Conference Series*, 1340(1), 012003. IOP Publishing.

Wang, L. H., Chen, B., Hwang, G. J., Guan, J. Q., & Wang, Y. Q. (2022). Effects of digital game-based STEM education on students' learning achievement: a meta-analysis. *International Journal of STEM Education*, *9*(1), 1-13.

Wang, L., & Chiang, F. K. (2020). Integrating novel engineering strategies into STEM education: APP design and an assessment of engineering-related attitudes. *British Journal of Educational Technology*, *51*(6), 1938-1959.

Zhao, Y. (2022). Systematic Analysis of Research Trends in STEAM/STEM Education Based on Big Data. In *2022 International Conference on Educational Innovation and Multimedia Technology (EIMT 2022)*,155-168. Atlantis Press.

Chapter 2

STEM Alignment for Industry 4.0: A Global Overview on Education 4.0

Suruchi Sharma[1,*]
and Nidhi Chaturvedi[2]
[1]School of Management, Graphic Era Hill University, Dehradun, Uttarakhand, India
[2]School of Business Management, Emirates Aviation University, Dubai, UAE

Abstract

The education sector has always been the interface between the process of knowledge assimilation and its real-world application. Throughout the world, universities, as seats of higher learning, have exhibited their sustained level of commitment towards the evolution of such educational methods and curricula that are capable of honing the future workforce with the requisite knowledge base, skills, and attitudes. Society, people, and economic development rest on such institutions for the holistic development of humankind. History stands as testimony that all the prior path-breaking industrial revolutions of the modern world set the pace and direction of the structure of education, which was metamorphosed following the knowledge required for real-world problems. The technological interface in the educational sector has been unprecedented in recent years. Technology has pervaded teaching across all levels, spanning from school to higher and professional education. Simultaneously, digitalization is transforming the domain skills and value sets of the existing industries and their workforce. Technological advances are evolving at a very fast pace. An important step towards digitizing production and building a cyber-physical system is Industry 4.0. This inevitable and forceful change is having an overwhelming

* Corresponding Author's Email: suruchisharma210876@gmail.com.

In: STEM
Editors: Priya Jindal, Amandeep Singh and Rashmi Aggarwal
ISBN: 979-8-89113-305-1

impact on the institutions of higher learning which are responsible and accountable for training the talent, leading scientific innovation from the front, disseminating know-how, and educating and skilling the future workforce. This has dramatically changed the living and working modes for individuals, societies, and countries. The digital revolution has brought to the forefront the necessity of adopting a novel educational model to cope with the destructive changes. The present chapter aims to provide an insight into the dimensions of Industry 4.0, Education 4.0, (encompassing STEM and its variants), and their inter-relationships to prepare global intellectuals equipped with the tools and techniques for requisite success in the world of disruptive technology. Contextual discussions to adapt and streamline the mammoth changes taking place are also addressed

Keywords: industry 4.0, education 4.0, STEM, workforce, innovation

Introduction

Human society has forged ahead from one transition to another starting from agriculture to industrial and recently to digital revolution. New vistas of technological disruption and digital metamorphosis are the global buzzword. The novel panorama is hailed as a mix of systems and methods that have diminished the traditional boundaries among the 'physical', 'digital', and 'biological' domains (World Economic Forum, 2018). The last few decades have seen technologically induced economic growth culminating in revolutionizing the lifestyles of people across the globe, though there are gaping differences due to many reasons. The organizations stand radically changed due to digital interventions, with refined and upgraded employment parameters. Therefore, the teaching-learning continuum also shifts to keep up with the changed scenario. The futuristic industry is slated to be 'smart', 'networked', and heightened mutual interaction between the physical and digital turfs. 'Smart human beings' and 'Smart machines' are expected to co-exist in the work area, yielding synergistic performance. There is no alternative for learning to co-exist. The extant modifications are deep-rooted enough to leave a profound impact on all the dimensions of the production process. Percolating from top to bottom, the entire value chain is being overhauled. The current revolution has mandated the use of multiple digital technologies across the industrial spectrum like manufacturing, research & development, ICT or information-communication-technological, educational, medical, agricultural and allied, food processing, etc. (Kumari et al., 2022). The uniqueness lies in the context of agility, which incorporates the smart usability of the manufacturing structure. The products are digitally integrated and efficiently communicate in real-time, analyze, and securely use the information to tackle future uncertainty. The 21st century has given us a dramatically changed, technologically transformed, and informationally overloaded world. We are now at the beginning of Industry 4.0, and this advancement revises life, business, education, jobs, skills, and attitudes. The natural extension of such transformation is a pressing requirement to build up the reigning educational and training setup for tackling the future-ready workforce.

The International Engineering Alliance's Washington Accord, which was established as a result of pertinent deliberations, was the first step towards officially recognizing engineering graduates and their degrees on a global scale. The parity expectation was a driving force to support the multinational and multicultural workforce across the globe. The accreditation system agreed upon by all members is implemented by the Washington Accord signatory nations. It has established a framework for the focus areas to be aligned for educators or lifelong learners from different disciplines of know-how. The 'Outcome Based Education' or the OBE framework provides the characteristics of "graduating engineers" (Felder and Brent, 2003). These outcomes have been frequently updated as per the requirements. The best educational program to address these issues concerning the industrial revolution 4.0 is "STEM" education which includes the domains of "science", "technology", "engineering", and "mathematics". STEM combines the four essential foundations of "Science (S)", "Technology (T)", "Engineering (E), and "Mathematics (M)" that are relevant for a lifelong learner. The relevance of the 'STEM' method lies in the coexistence of each discipline symbolized by this acronym. As a result, STEM gradually teaches students a methodology for using linked knowledge systems and concepts to deliver answers to real-world issues. Cultivation of "Higher Order Thinking Skills" (HOTS) required for Industrial 4.0, motivates the professionals to exercise their STEM knowledge base. The world is becoming bipolar in terms of defining the lines between high-skilled and low-skilled assignments. Therefore, it is necessary to improve STEM education for our industry 4.0-ready engineers.

Global Estimates for Industry 4.0, Education 4.0, and Employability

Although this radical change has been the subject of attention for some time, only a small number of governments have taken the policy-making and goal-oriented path. With a small number of recently industrialized nations, the majority are developed nations. Germany, the United States, Canada, Switzerland, Singapore, and Japan are among the developed nations that appear to be the most prepared and to be ahead of other newly industrialized nations like China, India, and so on. Although Thailand and other recently industrialized nations like it have good current manufacturing bases, it is felt that the process of future expansion could be difficult. An answer to Industry 4.0 is Education 4.0., which calls for collaboration between humans and technology to open up new possibilities. Although Industry 4.0 is one of the main strategies for nations to progress, each nation should have a unique approach to implementing the Education 4.0 plan (Hervas-Oliver et al., 2021). To say it differently, standardized plans and techniques for the educational revolution won't work for everyone or be appropriate for all nations. A significant majority of the populace worldwide is unaware of Industry 4.0 and how it will affect their lives (Tinmaz and Lee, 2019). With a lack of a suitable educational system or specific teaching and training programs in educational institutions, young generations in particular are being challenged to get ready for Industry 4.0. Examining students' readiness for the Industry 4.0 revolution is therefore crucial, especially for those enrolled in colleges or higher education institutions. Significant changes are anticipated as a result of this revolution, which will soon see these students make up the majority of the workforce in the nation.

It is well known that the workforce of the twenty-first century requires individuals who have relevant abilities. Fernandez-Feijoo et al. (2014) also highlighted the value of STEM education and some of its qualities that are important to business. They believed that the second decade of this era had started a global competitive pace in economic marketplaces, which initiated a shift in the instructional paradigm for teaching and learning. STEM education takes place in the real world; in this setting, schools collaborate with businesses and other educational institutions. According to a widely held belief, 65 percent of students presently studying or about to take admission in the primary set-up are eventually set to be professionally involved in the completely novel fields of work that are oblivious to as of now. According to

(Hoppe, 2016) strengthening and supporting entrepreneurship education is one of the EU's primary policy goals because it offers the knowledge, abilities, and mindsets necessary to foster an entrepreneurial culture. The new age of industrial revolution is all about the coexistence of digital and physical systems of working with unique communication and integration (Wang et al., 2016). Not all nations are moving in the same direction and at the same speed in this fast-paced climate.

Automation is occurring unevenly within countries as well as affecting different industries at varying rates. Over the past ten years, China has had the largest global expansion in the stock of industrial robots. 20% of all the robots being used in the industry come from this country, which is currently among the top five in the globe. It is also approaching a whooping world average of 70 robots per employee. However, there is significant variation between nations. In Korea, the top-ranked nation in the world, there are ten times more robots per 10,000 workers than there are globally. Industry 4.0 is also taking place in a setting where advanced economies are committed to putting measures in place to maintain their leadership. Chinese mainland potential and domestic demand are seen as China's new sources of demand and growth, and the country is aiming to capitalize on these opportunities. These developments are occurring as developing countries start to think about ways to boost their earnings from trade and investments as well as how to increase the domestic value added in their manufacturing processes. In this context, Industry 4.0 is a source of a drastic transition for the developing economies. The opportunity to lessen the stark disparity in development stages and close the development gaps at previously inconceivable rates and scales is being provided by emerging technologies. Digital technologies are not only fostering the expansion of start-ups in Silicon Valley but also in new ecosystems. By obtaining funding and exporting items internationally, they help innovative businesspeople all over the world convert their ideas into lucrative enterprises. Agriculture might become more environmentally and economically efficient, all credit to Industry 4.0. The importance of public-private partnerships in achieving this is rising. For instance, IBM has developed a platform that makes use of big data and artificial intelligence to assist farmers in making decisions. Due to the transformed business environment, there are both predictable and unforeseen changes in all spheres (Nathalie, 2018). On one hand, these robots will replace labor and be more productive because they can accomplish tasks faster and more accurately than humans (Pfeiffer, 2016). Contrarily, if this shift to a new industrial setup is successful, a reduction in the amount of manual labor and a change to positions that are more personally rewarding for workers may have a substantial positive effect on future working conditions (Riminucci, 2018). Robots replace labor-intensive jobs rather than a particular kind of work, as per the "International Federation of Robotics". Robots give workers better options to pursue more specialized, valuable, and well-paying occupations (IFR, 2022). Industry 4.0's potential impact on work prospects is under wraps still, and its net result will be greatly influenced by how the labor market operates and how firms, unions, and governmental organizations interact. According to Martin Ford's theory, it is proposed that automation, technologies, and robots are going to be involved in the labor-intensive jobs historically done by labor (Richardson, 2021). New industries will be created by creative destruction, and they are more likely to be technology-intensive than the old ones. The impact on employment and the future of work due to Industry 4.0 are being constantly debated by academicians and institutions between two opposing points of view. One says that if many current employments are replaced by cutting-edge technology and robotics, Industry 4.0 will result in significant job losses. On the other hand, many believe that Industry 4.0 would lead to the creation of new jobs and employment sectors, raising employment rates (Riminucci, 2018). 66.6% of manual jobs currently held in developing countries could be eliminated by automation and robotics (Grassi, 2020). A prediction states that while 45% of existing employment currently held by employees will likely be automated, 5% of full-time jobs will likely be replaced by technology (Manyika et al., 2017). However, some research asserts that persons with lower skill levels will be significantly more influenced, while others assert that people with medium ability levels will be more impacted (Jindal et al., 2016). On the side of the optimists, some studies claim that Industry 4.0 may enhance employment chances by creating new professions, especially in the disciplines of architecture, engineering, computer science, and mathematics. In reality, it has been asserted that historically, technological development has benefited employment. As an illustration, over the previous 15 years, the Internet has resulted in the loss of 500,000 employment in France, but it has also led to the creation of 1.2 million new jobs (Balliester and Elsheikhi, 2018). The deployment of Industry 4.0 technologies, according to some analysts, will not result in job losses even if it does not guarantee an increase in net employment. They believe that Industry 4.0 will lead to an improvement in worker productivity and product quality. Businesses will hire personnel with new credentials and skill sets, re-organize their organizational structure, and employ new management strategies to make better use of their resources (Kergroach, 2017). The qualities that employers will be looking for the most shortly include knowledge of information and communication technology (ICT), creativity, and decision-making abilities. Jobs that heavily rely on human skills, such as those in customer service, sales, marketing, training and development, people and culture, organizational development, and innovation management, are more likely to expand. Education cum training needs to be adapted for accommodating new credentials (Kergroach, 2017). Development of "STEM" is critical and simultaneously equal emphasis ought to be given to human-specific behavior patterns and skill sets that cannot be performed by digitized versions (Ackermann et al., 2014). Teamwork, multidisciplinary collaboration, and partner networks are more likely to become increasingly important, according to existing research on the practices of Industry 4.0 as they relate to human resources and workplace organization. Continuous change in the kinds of jobs people do and how they do them is nothing new. Governments, industries, and society have continuously failed to make important decisions about how to re-skill human resources for future workforces. We have a lot of things to learn from in today's transition, which is a significant distinction. Since the industrial revolution of the nineteenth century, a lot has been learned about how labor markets are changing. Employability in the knowledge economy is directly correlated with pertinent education and training. The knowledge economy is founded on information and knowledge, which creates an organization's intellectual capital (Benešová and Tupa, 2017) and propels economic growth and employment across industries (Dudersta et al., 2008).

Industry 4.0: Dimensions

The fourth industrial revolution, or Industry 4.0, is brought about by rapidly developing technologies like big data analytics, business intelligence, the Internet of Things, cloud computing advancements, cyber-physical systems, the internet and telecommunications industry, education, robotics, smart manufacturing, smart factories, autonomous systems, knowledge, innovation, and machine learning.

The following nine technologies have been linked by researchers to Industry 4.0: vertical and horizontal system integration, big data analytics, cloud computing, cyber security, augmented reality, autonomous robots, and additive manufacturing (Hizam and Soomro, 2021). The technologies associated with Industry 4.0 are shown as 'Dimensions of Industry 4.0' in Table 1. As a result, to flourish in Industry 4.0, the next generation must show that they have the necessary STEM competencies. The fourth industrial revolution is being fueled by several other reasons. Along with the idea of Industry 4.0, these variables also include market expansion, rising competitiveness, and globalization. Every industrial revolution resulted from a change from an established pattern to an inventive trend. For instance, the first industrial revolution came about as a result of the switch from manual to mechanized production, as well as the development of water and steam power. Similar to the third industrial revolution, the fourth is propelled by advances in machine-to-machine communications, autonomous systems, and smart manufacturing. Industry 4.0 demands a greater understanding of STEM capabilities.

Table 1. Dimensions of industry 4.0

S. No.	Technologies associated
1.	Augmented reality
2.	Autonomous robots
3.	Big data analytics
4.	Cloud computing
5.	Cyber-security
6.	Internet of things
7.	Simulation
8.	Vertical & horizontal system
9.	Additive monitoring

Source: Author's compilation.

To highlight the competitive strategy model of European Union businesses with other global markets, the phrase "Industry 4.0" was first introduced in 2011 at the Hanover Fair in Germany. It's critical that business managers comprehend Industry 4.0 and how it functions in practice. It is possible to construct a vision for Industry 4.0 that is unique to a plant by identifying the right model and having strong development support for it. The concept of Industry 4.0 needs to be connected to particular factory concerns. To move closer to Industry 4.0, it is vital to identify activity areas. The areas of activity that should receive priority could be a challenge. If the organization is out-of-date and the technology is modern, the system won't work correctly. The goal of Industry 4.0 is to develop intelligent goods, processes, and practices. Workers, machinery, and resources all have easy communication in the smart factory. The Internet of Things and Internet of Services, which refers to the pervasive connectedness of people, things, and machines, are the foundation of Industrial Vision 4.0. Products, vehicles, and tools "cooperate" to improve the next step in the production process. The virtual world and actual physical objects in the real world can connect that way. Smart items contain all of their information. They are aware of the dates and conditions of their manufacturing, the location of their stock, the estimated delivery date, and other crucial details. They can interact with networks, logistics, and smart mobility. Information delivery across the entire network (from the supplier to the client) is the goal of this integration. Suppliers can better plan and organize future deliveries because they are always kept up to date on the status of the goods through the integration that has been provided. Today, customers should contact the manufacturer to inquire about the status of their products; however, horizontal integration will allow customers to keep an eye on their products. If an issue arises, the consumer can act right away and choose how to solve it. To exchange information from the bottom of the hierarchy to the top and vice versa, vertical integration refers to the integration of information systems at various hierarchical levels. Manufacturers monitor the quality and cut down on waste by connecting all processes and utilizing sensors at every stage of manufacturing. When a possible problem is identified, the responsible personnel can take preventive action thanks to the continuous exchange of information through a hierarchy. Machines have access to information at all times since they are connected to a communication network. Businesses are aware of the benefits of Industry 4.0, including faster, smarter, and more sustainable productivity.

Education 4.0: STEM Enabled

21st-century skills are defined as "collaborating, communication, critical thinking, and creativity" (Partnership for the 21st Century Skills, 2014). The skills of the twenty-first century include "complex communication skills, self-development, systematic thinking, and non-routine problem solving". Innovation, employability, and effective teamwork can also be listed as 21st-century abilities in addition to these. As with most issues in the real world, STEM (science, technology, engineering, and math) is a synthesis that ties to daily life experiences and the workforce (Hafni et al., 2020).

To tackle an issue, one discipline is rarely sufficient; typically, it needs expertise from multiple disciplines. Therefore, it is more important than ever for STEM skills to be integrated into the classroom. STEM abilities emphasize mixing more than one field of science, technology, engineering, and mathematics to create educational materials that practically solve real-world issues. The students from such a setting will have no trouble understanding the intricate details of the issue, correlating those details to the relevant STEM field, and coming up with a workable solution. About Industry 4.0 and the constantly changing world, STEM skills in education 4.0 also put a priority on improving students' critical thinking skills. This empowers them to meet difficulties, break down boundaries, and become innovators and problem-solvers.

Education 4.0 is a teaching and learning approach that is compatible with Industry 4.0 and emphasizes the future transformation of education through automation and cutting-edge technology. Beyond only exploiting cutting-edge technologies in teaching and learning, Education 4.0 is a strategy for reforming education so that students can gain the skills they will need in the future. All students have access to it, routine administrative tasks are automated, students are building skills for the future workforce, instruction is individualized, and students receive helpful feedback. These are the five essential parts of this system. Table 2 displays these Education 4.0 components. The teaching and learning strategies need to change for educational institutions to develop the workforce needed for Industry 4.0. The students need to be equipped with knowledge and abilities that will be useful in future industries. In addition to preparing students for their careers, education should give them the skills necessary to compete in the constantly changing job market, teach them how to monetize their skills, give them the tools to improve their communication skills, form professional networks, and collaborate with others. A nation must improve its STEM skills system to meet the demands of Industry 4.0 if it wants to be a leader in the fourth industrial revolution.

Therefore, educational institutions must undergo a metamorphosis to develop the qualified workforce, thinkers, and innovators required to satisfy future demands. Jobs are getting more and more automated in today's globalized society where new technical developments occur daily. As a result, in all the industries that are expanding quickly, there is a growing need for highly qualified people with solid domain knowledge and technological abilities. In 2001, the "Scientific Directors - National Science Foundation" announced the abbreviation of 'STEM' (NSF Statutory Mission, 2002).

A STEM curriculum is designed to provide its students with the maximum possible practical exposure and hands-on experience in a particular field. Science lovers, who do not have any prior experience of such a learning environment, may experience this learning through higher education, that is, the bachelor's and master's level education.

Students who undergo STEM education develop a set of skills as described in Table 2 below:

In other words, STEM refers to experience-based learning that is structured around challenging technical group projects and aids in an individual's professional growth to help them excel in any technical field.

Table 2. Skills set: for 'STEM' graduates

Technical dimension	Personality dimension
Problem-solving	Decision making
Critical analysis	Management
Creativity & innovation	Drive & passion
Digital literacy	Communication skills (verbal, body language)
Research skills	Teamwork
Higher productivity	Leadership

Source: Author's compilation.

Table 3 given below enumerates the dimensions of 'Education 4.0' which is considered to be the best method to fill the skill gap to deal with the fourth industrial revolution IR 4.0, using the STEM and its variants.

Table 3. Dimensions of education 4.0

S. No.	Dimensions
1.	Accessible to all learners
2.	Automated basic administrative tasks
3.	Skilling learners for the future industries
4.	Personalized education
5.	Constructive criticism

Source: Author's compilation.

Students are given a solid academic and practical basis through their hands-on learning in real-world projects and workshops. The STEM education system's best quality is that it turns students into professionals who can adjust to a changing work environment. Table 4 depicts the dimensions of STEM education.

Table 4. Dimensions of STEM

S. No.	Dimensions of STEM
1.	Vocational training
2.	Experimental learning
3.	Real-world exposure
4.	Better networking opportunities
5.	More professional demand

Source: Author's compilation.

'STEAM' Education is seen as STEM education's successor since it combines the arts with STEM subjects including science, technology, engineering, and math. In technical domains, STEAM education aims to promote creativity, curiosity, and individual expression. It has a human touch thanks to the inclusion of the arts in STEM courses. Industry 4.0 is a complicated system that is difficult to implement. To realize and streamline its execution, it would reportedly take at least five years (Machado and Davim, 2020). A smart industry has many different aspects, thus there are now numerous obstacles to its employment, including political issues, economic and social issues, scientific and technological hurdles, etc. Sincere efforts to address these issues holistically would make implementation simpler and more organized. Undoubtedly, one of the biggest obstacles most nations face on their way to Industry 4.0 is related to the workforce's lack of the necessary skills. The higher education system or the universities must change in the quest to build critical abilities, intellect, and knowledge and to prepare the workforce for Industry 4.0.

The history of Education 4.0 dates from the 1.0–3.0 economic development era. The primary goal of education in version 1.0, according to the University of Phayao's Green assessment, was to provide human resources to support the agricultural economy and the traditional rural society. To operate the light industry, the country imported substitutes, used more natural resources, had a low cost of labor, and developed the nation by utilizing new technology. Education 2.0 was a revolution in education that focused on producing graduates for working in industry, especially light industry. To meet the demand for skilled human resources in heavy industry businesses and foreign direct investment, Education 3.0 has continuously improved the way technology is used. Global networks and the internet are now used to share and earn knowledge for improving learning skills. The fundamental goal of this educational age, known as "Education 4.0," is to generate graduates who have considerably greater skills in innovation and digital technology than those who graduated in the past. The ability to think creatively and innovatively is important for overcoming a country's poverty and achieving economic and social sustainability. STEM knowledge, thinking skills, multi-literacy, and socio-emotional intelligence are the four key categories of STEM capabilities that underpin their utility in achieving the necessary skilled workforce for Industry 4.0. For instance, STEAM promotes teamwork to comprehend a STEM idea. STREAM (Science, Technology, Reading and Writing, Engineering, Arts, and Mathematics) and STEAM (Science, Technology, Engineering, Art, and Mathematics) include a wider spectrum of educational opportunities for learners. Information and communication technologies are used in STEM education as a teaching tool and as a subject of study.

The future generation of workers will require competency in communication, collaboration, skills for digital transformation, solving complex problems, and leadership (Akgunduz and Mesutoglu, 2021). Successful transition to work-life necessitates students' familiarity with STEM disciplines and Industry 4.0 components.

First, STEM instruction focuses on issues and challenges that exist in the real world. Students in STEM courses face and seek solutions to current social, economic, and environmental problems. The engineering design process (EDP), which is a second differentiating characteristic, guides STEM classes. The EDP provides a method that is adaptable and walks students through identifying an issue or a design challenge, as well as coming up with and putting into practice a solution. The third attribute of STEM instruction is that it engages students in exploratory, open-ended research. In this characteristic, students' work is practical and collaborative, and decisions about the best solutions are based on their input. The ability to communicate allows students to express their ideas and, if necessary, revise their prototypes. Their own thoughts and research are within their control. The fourth attribute, that STEM education fosters, involves students in cooperative learning. The fifth characteristic is that STEM programs necessitate that students understand challenging scientific and mathematical concepts, who may then understand that science and math are not distinct subjects because their education immerses them in both and calls for teamwork to solve problems. The last argument is that STEM education encourages many accurate responses and reinterprets failure as an essential component of the learning process. Problem-solving skills include trouble-shooting situational deadlocks, observation, and inquiry. It shows up as a part of 21st-century abilities including creativity, critical thinking, and problem-solving.

According to a widely held belief, it was predicted that 65% of today's kindergarteners will ultimately wind up working in entirely new job types that have not yet been formed. One of the main aims of the STEM program, which provides the abilities, information, and mindsets required for an entrepreneurial culture, is to enhance and promote a varied curriculum.

Global Response and Challenges

There is no stopping the rapid technological progress we are currently seeing. We have made rapid advancements in technology, from the development of the steam engine to the creation of the first smart factory. We are currently moving in a direction that effectively improves the quality of human habitation as we experience Industry 4.0. To keep up with the recent tremendous changes in technology, we have not made enough progress in the way we teach. The most effective way to prepare people is through education. Our educational system needs to change to keep up with the countless technological developments that are taking place around us. For students to effectively work towards developing new ideas, we need an educational system that fosters creativity. Such a system must instill pupils with faith and confidence.

Industry 4.0 has the potential to transform emerging economies. It is not a free pass for their advancement, though. While some run the risk of ever-increasing marginalization, certain developing economies run the risk of losing the manufacturing capacity and jobs they have earned through diligent work. The outcome will depend on several variables, but especially on the ability of governments, businesses, and the various societal groups in developing economies to recognize the magnitude of the potential impact of Industry 4.0 on their development and in the strategies that they will define and implement. For emerging nations to gain from Industry 4.0, quick action and a clear grasp of the societies they want to become are essential. Without clear goals, principles, and options, Industry 4.0 runs the possibility of becoming a brand-new, much more significant factor in regional and global development disparities. The developing world's economies must confront this transformation while expressing their own hopes and worries. Developing nations must be aggressive on several fronts. How to unleash Industry 4.0's development potential must be on the national agenda. They require more than just forward-thinking administrations (and leaders) to do this. To make the shift easier, they also require access to expertise and technical assistance. Despite structural issues that make leapfrogging to Industry 4.0 a reality difficult for developing nations to achieve, emerging economies have a significant advantage. According to history, reforms are more likely to be implemented when there is a desire for change rather than a desire to maintain an existing status quo.

The Asian Tiger Economies—China, Malaysia, India, and Singapore—have placed a strong emphasis on providing STEM programs in their educational institutions as they work to develop their systems. Over the past three decades, China in particular has seen a spectacular rise in the number of STEM graduates, giving the nation a skilled and plentiful workforce in science and technology. A further indication that STEM education is promoted and valued is the fact that 31% of bachelor's degrees in China are conferred in engineering, according to the National Science Foundation. Forbes has also acknowledged that China's drive for STEM-educated human capital has resulted in it being a world-class leader in developing tech disciplines and manufacturing advances, thereby enhancing its competitiveness and carving out a position in the global market.

Developing countries must understand the potential of Industry 4.0. To fully realize the potential of digital technologies and lessen the risks of losing money as a result of reshoring, the countries that currently have manufacturing plants, such as Mexico, Malaysia, and Vietnam, must figure out how to modernize their economies. Other countries that never experienced true industrialization will need to do so in an Industry 4.0 context. Although these economies are still far from having hyper-connected smart sectors, governments and businesses must now consider these realities in their strategy to avoid more marginalization in the future.

Stakeholders

Development that is inclusive of all parties involved, including business, academics, the government, society, and international organizations is sustainable. To improve STEM education, it is necessary to boost the roles of the nation's workforce, which comprises the educational system, culture, and government. This is crucial to foster and promote STEM enthusiasm in the young generation. The roles of the national workforce, which include the education system, culture, and government, are crucial in creating the human capital armed with education, knowledge, experience, skills, creativity, and motivation. The government's job is to offer a support network, particularly in the form of financial aid for students majoring in STEM subjects. As the market salary supplied by the industry still does not outweigh the burden of the cost of studying, and as the salary range is somewhat equivalent to the income of non-STEM groups, financial aid could cut the cost of education and boost return on investment. Finding the best teaching method for STEM disciplines inside an institution may be the deciding factor. Even though many policies have been created, teachers must still use their creativity and interpretation to carry out the policies at the school level. Parents must intuitively and overtly value science education for the STEM fields to develop in society. Children can select their areas of interest based on their experiences and the objective is to build a STEM field setting that is valuable. Another substantial difficulty that needs to be addressed by the concerned parties is the training and preparation of the STEM tutors. The growing technical divide could limit developing nations' ability to catch up, which would raise wealth inequality across nations if governments do not take deliberate action to exploit the opportunities presented by this new wave of technological progress. In the end, what we should be looking for is not Industry 4.0 as the ultimate goal, but rather, strategic methods to use it to build our economies more quickly, inclusively, and in an environmentally friendly way.

Conclusion

The need for countries and organizations to implement education systems that are more focused on knowledge beyond what the world now preaches has become vital as people embrace Industry 4.0. It might be necessary to start teaching creativity to kids at a young age and continue through college. The traditional educational methods of writing, reading, cramming, and memorization as a means of passing an exam, which never generate thinkers, creators, or ingenuity, should be abandoned. Nations must therefore transform their educational institutions to create superhuman capable of surviving the Industry 4.0 revolution. To achieve the goals of the Industry 4.0 revolution, a country's culture must be conducive to such government initiatives and where people feel they can make a difference (Agolla, 2018). To maintain robust economic growth and encourage investment in a nation, the educational system must transform. The market arrangements for education will sculpt consumer behavior that affects productivity and attainment. Utilizing Industry 4.0 technology in the manufacturing sector increases productivity, lessens the environmental damage caused by industrialization, and generates new jobs rather than job losses. However, Industry 4.0 offers developing nations, who may have missed the early changes brought about by industrialization, a chance to compete on a new level. As a result, developing nations cannot afford to ignore this new wave of technological progress. This transition will be largely dependent on new national strategies, policies, legislation, programs, and activities to encourage and facilitate the change. To get people and industry ready for a time of rapid and disruptive change, each nation will need science, technology, and innovation policies appropriate to its degree of development. The Fourth Industrial Revolution requires a wide range of skills, including those related to coding, cloud computing, artificial intelligence, the Internet of things, blockchain, robotic programming, data science (big data & analytics), cyber security, and so on. Outcome-based education must teach these skills starting in early childhood and continuing through the tertiary level. Thus, re-skilling and up-skilling will be necessary, but they call for a systematic strategy to create a strong and diverse human capital to fuel the Industry 4.0 transition. Since technological advancement is an unavoidable part of life, Industry 4.0 won't be the last in its queue. The greatest approach to do this is through employing Artificial Intelligence in schools to educate children for the technological future that awaits them. Keeping all the stakeholders in a loop is also a fundamental factor in its success.

References

Agolla, J. E. (2018). Human capital in the smart manufacturing and industry 4.0 revolution. *Digital transformation in smart manufacturing*, 41-58.

Akgunduz, D., and Mesutoglu, C. (2021). Science, Technology, Engineering, and Mathematics Education for Industry 4.0 in Technical and Vocational High Schools: Investigation of Teacher Professional Development. *Science Education International*, 32(2), 172-181.

Ackermann, H., Hage, S. R., and Ziegler, W. (2014). Brain mechanisms of acoustic communication in humans and nonhuman primates: An evolutionary perspective. *Behavioral and Brain Sciences*, 37(6), 529-546.

Benešová, A., and Tupa, J. (2017). Requirements for education and qualification of people in Industry 4.0. *Procedia manufacturing*, 11, 2195-2202.

Pfeiffer, S. (2016). Robots, Industry 4.0 and humans, or why assembly work is more than routine work. *Societies*, 6(2), 16.

Balliester, T., and Elsheikhi, A. (2018). The future of work: a literature review. *ILO Research Department Working Paper*, 29, 1-54.

Felder, R. M., and Brent, R. (2003). Designing and teaching courses to satisfy the ABET engineering criteria. *Journal of Engineering education*, 92(1), 7-25.

Fernandez-Feijoo, B., Romero, S., and Ruiz, S. (2014). Effect of stakeholders' pressure on transparency of sustainability reports within the GRI framework. *Journal of business ethics*, 122, 53-63.

Grassi, A., Guizzi, G., Santillo, L. C., and Vespoli, S. (2020). The manufacturing planning and control system: A journey towards the new perspectives in industry 4.0 architectures. *Scheduling in Industry 4.0 and Cloud Manufacturing*, 193-216.

Hafni, R. N., Herman, T., Nurlaelah, E., and Mustikasari, L. (2020). The importance of science, technology, engineering, and mathematics (STEM) education to enhance students' critical thinking skill in facing the industry 4.0. In *Journal of Physics: Conference Series*, 1521 (4), 042040. IOP Publishing.

Hizam-Hanafiah, M., and Soomro, M. A. (2021). The situation of technology companies in industry 4.0 and the open innovation. *Journal of open innovation: technology, market, and complexity*, 7(1), 34.

Hervas-Oliver, J. L., Gonzalez-Alcaide, G., Rojas-Alvarado, R., and Monto-Mompo, S. (2021). Emerging regional innovation policies for industry 4.0: analyzing the digital innovation hub program in European regions. *Competitiveness Review: An International Business Journal*, 31(1), 106-129.

Hoppe, M. (2016). Policy and entrepreneurship education. *Small Business Economics*, 46(1), 13-29.

IFR.(2022). *International Federation of Robotics*. Retrieved from https://ifr.org/downloads/press2018/2022_WR_extended_version.pdf.

Jindal, P., and Tej, F. (2016). Being an educator in private vs. public school system: A comparative study on job contentment. *IRA International Journal of Education and Multidisciplinary Studies*, 3(3), 309-319.

Kumari, S., Jindal, P., and Chand, P. K. (2022). Social Media Networking: As Facilitator of Acculturation. *ECS Transactions*, 107(1), 9381.

Kergroach, S. (2017). Industry 4.0: New challenges and opportunities for the labour market., *Foresight and STI Governance*, 11(4), 6-8.

Machado, C., and Davim, J. P. (Eds.). (2020). *Industry 4.0: Challenges, trends, and solutions in management and engineering*. CRC press.

Manyika, J., Lund, S., Chui, M., Bughin, J., Woetzel, J., Batra, P., and Sanghvi, S. (2017). Jobs lost, jobs gained: Workforce transitions in a time of automation. *McKinsey Global Institute*, 150(1), 1-148.

Nathalie Udo.(2018). An *Agile Mindset For Today's Complex and Unpredictable World* Retrieved from https://www.linkedin.com/pulse/agile-mindset-todays-complex-unpredictable-world-udo-mba-pmp on October 26, 2018.

National Science Foundation. (2002). Retrieved March 29, 2002, from https://www.nsf.gov/pubs/2002/nsf02105/nsf02105.pdf.

Partnership for the 21st Century Skills. (2014). Retrieved September 18, 2014, from https://feilidi.weebly.com /uploads/ 4/2/5/8/42583553/p21_skills.pdf.

Riminucci, M. (2018). Industry 4.0 and human resources development: A view from Japan. *E-Journal of International and Comparative Labour Studies*, 7 (1), 6-15.

Richardson, R. (2021). Defining and demystifying automated decision systems. *Md. L. Rev.*, 81, 785.

Tinmaz, H., and Jin, H. W. A. (2019). A preliminary analysis on Korean university students' readiness level for Industry 4.0 revolution. *Participatory Educational Research*, 6(1), 70-83.

Wang, S., Wan, J., Li, D., and Zhang, C. (2016). Implementing smart factory of industrie 4.0: an outlook. *International journal of distributed sensor networks*, 12(1), 3159805.

World Economic Forum. (2018). Retrieved January 28, 2018, from https://www.weforum.org/whitepapers/eight-futures-of-work-scenarios-and-their-implications.

Chapter 3

A Systematic Literature Review of STEM Education in Changing National and Global Policies and Procedures

Madhavi Shamkuwar[1,*], PhD
Pandurang Patil[1]
Rahul More[2], PhD
and Sujata Barde[3]
[1]Zeal Institute of Business Administration, Computer Application & Research, Savitribai Phule Pune University, Pune, India
[2]Sinhgad Institute of Management & Computer Application, Savitribai Phule Pune University, Pune, India
[3]Pune Institute of Computer Technology, Savitribai Phule Pune University, Pune, India

Abstract

All over the world, STEM education is an education revolutionary movement that has gained a lot of momentum in the last few decades. "STEM" education follows a transversal education approach and it is a focused curriculum placing a lot of emphasis on Science, Technology, Engineering, and Maths. All these subjects/courses are taught to develop creativity, critical & logical thinking skills in a student. These important academic areas are approached by STEM schools and programs in a way that makes connections between them. Depending on the resources available in a particular school district or location, STEM-focused learning programs are available from preschool through master's degree

* Corresponding Author's Email: madhavi.sh@gmail.com.

In: STEM
Editors: Priya Jindal, Amandeep Singh and Rashmi Aggarwal
ISBN: 979-8-89113-305-1

programs. STEM-based learning initiatives are designed to spur students' interest in pursuing employment in those sectors and higher education. A blended learning approach is frequently used in STEM education, which blends traditional classroom instruction with online learning and hands-on experiences. The abilities acquired through STEM education go beyond those required to succeed in STEM disciplines, enabling youngsters with a variety of interests to enter any business and succeed by equipping them with useful skill sets enabling future generations' career success.

Keywords: STEM education, artificial intelligence, bibliometric analysis, virtual reality

Introduction

Students must advance their STEM skills well beyond what was previously thought to be acceptable if they are to prosper in this new information-based and highly technological culture.

Students can develop solutions utilizing their knowledge and skills by participating in STEM education, which encourages the application of the design process. However, nowadays with the very disruptive nature of technology and due to the sudden outbreak of the pandemic schools, colleges, and universities across the globe have started augmenting STEM with the support of technology. Artificial intelligence, Data Science, Virtual reality, and the education enablers of STEM education.

Media and technology literacy, productivity, social skills, communication, adaptability, and initiative are examples of 21st-century abilities (Tytler, 2020). Problem-solving, critical thinking, creativity, curiosity, decision-making, leadership, entrepreneurship, acceptance of failure, and other abilities can be learned through STEM education. These skill sets go a great way toward training these kids to be innovative, regardless of the future professional route they are thinking about. Since governments are under increasing pressure to create technically trained workers from the obligatory education sectors, people able to actively participate in the so-called "knowledge economy" or knowledge society (Zouda et al., 2023), effective STEM education is of significant relevance on a global scale.

Reviews of the state and trends in STEM education research internationally support the development of the discipline due to the sharp rise in academic and research publications on the subject in recent years. This review is being conducted to make sure that technology is an enabler to STEM education to face the difficulties of a world that is changing, especially in this century, in the 21st.

This book chapter offers a global perspective on the impact of Artificial intelligence on STEM change. The book's use of local experts from a highly diverse cohort around the world to draw from their actual experiences and research is a significant and distinctive aspect of the work. The article introduces STEM education and its philosophy, National Education policy-2020, the 2030 Agenda for UN Sustainable Development adopted for quality education, Education 4.0, and associated challenges.

STEM Education in India

STEM education in India refers to education in the fields of Science, Technology, Engineering, and Mathematics. It has been a priority in India's education policy, with a focus on developing skills in these areas to meet the demands of a rapidly growing economy and technology-driven world. In India, STEM education starts at the primary and secondary school levels, with a curriculum that includes subjects such as mathematics, physics, chemistry, biology, and computer science. At the higher education level, there are numerous institutions offering degrees in STEM fields, including engineering, technology, and scientific research institutions.

Despite significant progress, there are still challenges in STEM education in India, including a lack of qualified teachers, limited access to resources and technology, and a need for more hands-on, project-based learning experiences. To address these challenges, the government and private organizations have implemented initiatives to improve STEM education, such as teacher training programs, scholarships for students, and funding for research and development. Overall, STEM education in India continues to play a critical role in shaping the future of the country and its economy, and efforts are being made to ensure that students have access to quality education and resources in these fields (Jindal et al., 2016).

The Indian government is keeping no stone unturned concerning education. The number of initiatives to foster a knowledge economy among young and budding professionals is a remarkable contribution. An innovative program called Atal Tinkering Lab has been launched by the Indian government (ATL). The fundamental goal of this program is to foster in Indian students a scientific temperament, originality, and creativity. This program aims to cultivate in young minds a sense of wonder, imagination, and creativity while also teaching them skills like design mentality, computational thinking, adaptive learning, physical computing, etc. Students have the opportunity to use tools and equipment through this program to better grasp the methods and components of STEM fields (STEM).

Numerous private businesses have taken the initiative to offer STEM education services to youngsters in our nation. More than one million students now have better access to quality STEM education, and 7,000 teachers now have access to improved methods for teaching science and mathematics, according to a report from the private organization STEM Building Brains - Beyond Books. TryScience has reached out to nearly 30,000 people in India. 1,44,000 classrooms are affected by 11,03,643 pupils, educators, and science teachers. With the help of five implementing partners and science-based NGOs, STEM pedagogy has been implemented in 9 states and 5 indigenous languages (Hindi, Tamil, Kannada, Gujarati, and Marathi) in India.

Children with special needs should receive STEM education since it is a fundamental human right. Special needs students should receive the same free, suitable education as other kids. A novel method of educating children with exceptional needs alongside their typically developing peers is known as inclusive education. No matter their abilities or flaws, it treats every pupil equally. With the appropriate adaptations or modifications in the following areas, such as content, presentation, and evaluation, the National Curriculum Framework for School Education has suggested inclusive schools for students with special educational needs (Wirzal and Putra, 2022).

STEM Education across the Globe

In many developed countries, STEM education starts at an early age with a focus on developing core skills and knowledge in these fields. At the higher education level, there is a wide range of programs and institutions offering degrees in STEM fields, with a focus on preparing students for careers in science, technology, engineering, and mathematics-related industries.

However, there are also challenges facing STEM education globally. One of the major challenges is a shortage of qualified teachers and instructors, particularly in developing countries (Ostreika et al., 2020). Additionally, there are issues with student engagement, with many students losing interest in STEM subjects as they progress through their education. To address these challenges, governments and private organizations are investing in teacher training programs, student outreach, and engagement initiatives, and scholarships and funding for research and development (Cheng, 2022; Avdeenko et al., 2021; Ada, 2016).

The use of STEM education is meant to improve human activities (Awosika, 2008). STEM has significantly changed economic life and has been the main driver of globalization. It was also noted that STEM has been acknowledged globally as a pre-requisite in technological development and possible ways to address developmental challenges such as environmental degradation, food insecurity, infant mortality, and job insecurity among others (Obianuju et al., 2013) This is important for any nation like Nigeria to achieve sustainable development. One of the goals of STEM education is this.

STEM education is expected to help the students develop into more responsible citizens who will help to build a strong economy contribute to a healthier environment and bring about a brighter future for everyone. STEM education could also help bring about a stronger society that is able to address some of the developmental challenges like environmental degradation and food insecurity. The study further stressed that STEM education helps build a strong thereby addressing the issue of poverty. STEM education is also expected to help produce more responsible citizens and help produce future parents who will have the capacity to produce safe, healthy environments for their own children that means that STEM education can help address social vices like militancy, poor health care, insecurity, and religious intolerance among others. STEM education is supposed to help contribute to longer, healthier, and better lives, and a healthier society means a more productive society (Nnorom, 2016). However, the present reality shows that the impact of STEM in addressing developmental challenges has not been felt in Nigeria.

Overall, STEM education plays a critical role in shaping the future of economies and societies, and efforts are being made globally to ensure that students have access to quality education and resources in these fields. The philosophy of STEM education (Science, Technology, Engineering, and Mathematics) is centered around providing students with a comprehensive education that prepares them for the challenges and opportunities of the 21st century. It is based on the belief that a strong foundation in STEM subjects is essential for success in an increasingly technological and scientifically oriented world (Kumari et al., 2022).

In a quest to understand the significance of STEM education, an advanced search was performed on J-Gate within a stipulated period of 2000-2023. As a result, the outcome was based on country-wise and subject-wise research trends. It was found that the top 5 countries concerning STEM education research trends are: the United States of America, Germany, the United Kingdom, Turkey, and Switzerland as shown in Table 1. Table 2 shows the subjectwise STEM education research trends these are education & careers, sociology, behavioural science, applied mathematics, and society and culture.

Table 1. Countrywise article published

Country	No. of articles published
United States of America	720
Germany	402
United Kingdom	333
Turkey	111
Switzerland	106

Source: Author's compilation.

Table 2. Subjectwise article published on STEM education research trends

Subject category	No. of articles published
Education & careers	1672
Sociology	275
Behavioural science	164
Applied mathematics	118
Society and culture	106
Business management	72
Psychiatry & psychology	72
Language & linguistic	70
Philosophy	69
Economics	66

Source: Author's compilation.

Trends and Key Principles

Some of the common trends in global STEM education include an emphasis on hands-on, project-based learning experiences to foster problem-solving skills and creativity (Darmawansah et al., 2023). It further considers investment in teacher training and professional development to ensure that teachers have the knowledge and skills to effectively teach STEM subjects (Szczotka and Szewczuk, 2021). The collaboration between government, industry, and academia supports research and innovation in STEM fields (Lee and Jung, 2021). Exploring competing perspectives on how to design an open innovation program for high school STEM education. Encouragement of diversity and inclusiveness in STEM education, with a focus on attracting and retaining underrepresented groups, such as women and minorities are prime important in STEM (Marshall et al., 2022). It also promotes the integration of technology and digital tools to enhance the teaching and learning of STEM subjects.

Some of the key principles of the philosophy of STEM education include:

i. *Problem-solving:* STEM education emphasizes the development of critical thinking, problem-solving skills, and creativity through hands-on, project-based learning experiences.
ii. *Interdisciplinary approach:* STEM education recognizes the interconnected nature of science, technology, engineering, and mathematics and promotes an interdisciplinary approach to learning.
iii. *Relevance:* STEM education aims to make the subjects relevant to students' lives and experiences, by connecting them to real-world problems and applications.
iv. *Collaboration:* STEM education values collaboration and teamwork, recognizing the importance of working together to find solutions to complex problems.
v. *Equity and Access:* STEM education recognizes the importance of providing all students, regardless of their background, with access to high-quality STEM education, and actively works to remove barriers to entry.

STEM Education and National Education Policy 2020

The NEP aims to transform India's education system and make it more relevant and accessible to students, with a focus on quality and equity. Some of the key provisions of the NEP 2020 relevant to STEM education include:

i. *Integration of STEM subjects into the curriculum:* The NEP 2020 emphasizes the integration of STEM subjects into the curriculum at all levels, from primary to higher education, with a focus on hands-on, project-based learning (Sharma et al., 2021).

ii. *Teacher training:* The NEP 2020 recognizes the importance of high-quality teacher training, and provides ongoing professional development opportunities for teachers in STEM subjects (Jha, and Parvati, 2020).

iii. *Encouraging innovation:* The NEP 2020 places a strong emphasis on promoting innovation and research in STEM fields, and provides for greater collaboration between government, industry, and academia (Shashidharan et al.,2021).

iv. *Access and equity:* The NEP 2020 recognizes the importance of providing access to high-quality STEM education to all students, regardless of their background, and works to remove barriers to entry (Panigrahi and Gupta, 2021).

Overall, the NEP 2020 recognizes the central role that STEM education plays in shaping the future of India and its economy, and aims to ensure that students have access to quality education and resources in these fields. The NEP 2020 is a significant step forward in India's efforts to strengthen its position as a leader in STEM education and research. The NEP 2020 represents a significant step forward for STEM education in India, with a focus on quality, equity, and relevance. Implementing these provisions will require collaboration between the government, industry, and academia, as well as sustained investment in STEM education.

STEM Education and UN Sustainable Goals

STEM education (Science, Technology, Engineering, and Mathematics) plays an important role in achieving the United Nations' Sustainable Development Goals (SDGs). The SDGs are a set of 17 goals aimed at promoting a more sustainable and equitable world, and STEM education can contribute to achieving several of these goals, including:

i. *Quality Education (SDG 4):* STEM education provides students with the knowledge and skills they need to succeed in a rapidly changing world, contributing to the goal of ensuring inclusive and equitable quality education and promoting lifelong learning opportunities for all (Maruf et al., 2022; Ling et al., 2019; Dentinho,2014).

ii. *Sustainable Cities and Communities (SDG 11):* STEM education plays a crucial role in developing the technological and scientific solutions needed to build sustainable cities and communities, such as renewable energy systems and sustainable transportation systems (Kufeoglu, 2022).

iii. *Responsible Consumption and Production (SDG 12):* STEM education helps to foster a deeper understanding of the science behind sustainable consumption and production, and supports the development of new technologies and systems to promote sustainable resource use (Kufeoglu, 2022).

iv. *Climate Action (SDG 13):* STEM education is essential for addressing the challenges of climate change, as it provides students with the knowledge and skills they need to develop and implement solutions to mitigate and adapt to the impacts of climate change (Roy and Ayalon,2022).

v. *Reduced Inequalities (SDG 10):* STEM education can help to reduce inequalities by providing access to quality education and employment opportunities in the growing STEM fields, particularly for underrepresented groups, such as women and minorities (Dirrler,2020).

In conclusion, STEM education is a critical tool for achieving the United Nations' Sustainable Development Goals, and investing in STEM education is essential for promoting a more sustainable and equitable world.

Education 4.0

Education 4.0 refers to the integration of technology and digital tools into the education sector, intending to create a more personalized, flexible, and accessible learning experience for students. STEM education (Science, Technology, Engineering, and Mathematics) plays a crucial role in Education 4.0 as it provides students with the knowledge and skills they need to thrive in an increasingly technological world. STEM education (Science, Technology, Engineering, and Mathematics) is a key component of Education 4.0, as these subjects are at the forefront of technological and scientific advancement (Holford et al., 2023; Kim, 2022).

Some of the key ways STEM education aligns with Education 4.0 include:

i. *Hands-on, project-based learning:* STEM education emphasizes hands-on, project-based learning, which is well-suited for the digital tools and technology of Education 4.0.
ii. *Personalized learning:* STEM education can be tailored to meet the individual needs and interests of each student, allowing for a more personalized and engaging learning experience.
iii. *Access to technology and resources:* STEM education integrates technology and digital resources into the curriculum, providing students with access to the tools they need to succeed in the 21st century.
iv. *Collaboration and teamwork:* STEM education values collaboration and teamwork, which can be facilitated through digital tools and technology in Education 4.0.
v. *Relevance to real-world problems:* STEM education connects students to real-world problems and applications, and technology can be used to make this connection even stronger.

In summary, STEM education is well-aligned with Education 4.0, as both seek to provide students with the knowledge and skills they need to thrive in a rapidly changing world. By integrating technology and digital tools into STEM education, we can create a more personalized, flexible, and accessible learning experience for students. In conclusion, STEM education is well-suited to the Education 4.0 framework, and the integration of technology and digital innovation has the potential to transform the teaching and learning of these subjects, promoting greater access, equity, and quality in STEM education.

Bibliometric Analysis

Science Direct is a platform for peer-reviewed academic journals and research articles. It is owned by Elsevier, a global information analytics business specializing in science and health. Science Direct provides access to over 15 million articles from more than 4,000 academic journals, covering a wide range of scientific, technical, and medical topics. It is widely used by researchers, academics, and students as a primary source of information for their work and studies. The platform also features tools for searching, browsing, and retrieving articles, as well as features to enhance the research process, such as citation and reference management tools.

Bibliometric analysis is the study and evaluation of academic literature and research output through the use of quantitative methods. It involves analyzing bibliographic data such as publication counts, citation counts, authorship patterns, and other metrics to gain insights into the structure and evolution of a particular field of research. Bibliometric analysis can be used to assess the impact of individual researchers, institutions, and countries, as well as the distribution of funding and research collaborations.

Bibliometric analysis is performed on the research articles considered for the current study. For the same 4062 papers were being considered for the systematic literature review and bibliometric analysis. In ScienceDirect Journal, an advanced search was performed. For the same two search keywords were given ‘stem education’ and ‘artificial intelligence’ and the year range was from 2000 to 2023. As a result of the search, 4062 distinct articles were generated for study. These 162 documents were extracted in different formats such as .ris, .txt, and .bio. To perform bibliometric analysis, VoSViewer is used to create different visualizations.

VOSviewer is a free, open-source software tool for visualizing and analyzing bibliometric networks. It allows users to construct and analyze co-citation and co-occurrence networks based on bibliographic data such as publication counts, citation counts, and co-authorship patterns. VOSviewer can be used to analyze a wide range of bibliometric data, including data from scientific publications, patents, and web-based sources. The software allows users to visualize the relationships between authors, journals, institutions, and keywords in a network format, providing an intuitive and interactive way to explore complex bibliometric data.

Table 3. Bibliometric analysis -full counting, minimum no. of occurrence of keywords = 3

Color	Keywords
Red	State, work, ability, resource, information, context, organization, time, implementation, control, article, problem, system, problem, user
Green	Machine learning, year, application, field, development, technique, approach, paradigm, challenge
Blue	Process, role, concept, artificial intelligence, explore, example, mean

Source: Vosviewer.

Furthermore, on these 162 documents, binary counting is being performed and related visualizations are being created. Binary counting in bibliometric analysis refers to the practice of counting a publication or a citation only once, regardless of the number of times it appears in a dataset. Binary counting is used in many bibliometric studies as a simplifying assumption, as it reduces the complexity of the data and allows for more straightforward analysis.

As depicted in Table 3, three distinct clusters are being created with a major focus on the area of interest 'entrepreneur', the minimum number of occurrences is 3. As a result of this 34 different documents are been considered and then three different clusters Red, Green, and Blue are generated as bibliometric analysis as shown in Table 3.

Applications of AI in STEM Education

The following are some common applications of AI in STEM education

1. *Adaptive learning:* AI can be used to personalize learning experiences by adjusting the difficulty level and content based on a student's performance (Zhong, 2018).
2. *Automated grading and assessment:* AI algorithms can be used to grade student assignments and assessments, reducing the workload of teachers and providing more efficient feedback (Fleming et al., 2010).
3. *Intelligent tutoring systems:* AI can be used to provide real-time, personalized support to students as they work through challenging STEM concepts (Hughes, 2021).
4. *Predictive analytics:* AI can be used to analyze data on student performance and provide insights into areas where students may need additional support. Virtual labs and simulations: AI can be used to create interactive virtual labs and simulations, providing students with hands-on experiences in a safe and controlled environment.
5. *Virtual teaching assistants:* AI can be used to provide additional support for students outside of class time, answering questions and helping students through difficult concepts (Levin et al., 2022).

Conclusion

In conclusion, STEM education is crucial for preparing students for careers in fields related to science, technology, engineering, and mathematics. It emphasizes critical thinking, problem-solving, and hands-on learning, providing students with the skills they need to succeed in the fast-paced and constantly evolving technology landscape. With a strong STEM education, students are equipped to tackle real-world challenges and drive innovation. Investing in STEM education is essential for both individual and societal growth and development.

STEM education is crucial in preparing individuals for the demands of a rapidly changing, technology-driven world. It provides students with the skills and knowledge necessary to succeed in a variety of fields, including science, technology, engineering, and mathematics. Additionally, STEM education helps to foster critical thinking, problem-solving, and creativity, which are essential qualities in today's workplace. However, despite its many benefits, access to STEM education remains unequal, particularly for underrepresented groups. It is therefore important for educators, policymakers, and communities to work together to ensure that all students have access to high-quality STEM education and opportunities.

References

Avdeenko, A., Sabirova, F., & Konyushenko, S. (2021). STEM Education: Key Challenges in the Common Practice. *International Journal of Web-Based Learning and Teaching Technologies (IJWLTT)*, *16*(6), 1-11.

Ada, N. A. (2016). The Dilemma of the Science Teacher in the Implementation of Science Education curriculum Reforms at the Basic Education Level in Nigeria. *BSU Inaugural lecture series (10).*

Awosika, I. F. (2008). The contributions of biological science to national development: Nigerian experience. *Ife Journal of Science*, *10*(1), 97-101.

Cheng, M. M. H. (2022). An Overview of STEM Education in Asia. *Concepts and Practices of STEM Education in Asia*, 1-15.

Dentinho, T. P. (2014). Urbanisation and sustainability: the role of the spatial allocation of property rights and public investment. *International Journal of Global Environmental Issues*, *13*(2-4), 189-205.

Dirrler, P. (2020). The Effects of Diversity on Achieving Reduced Inequalities. *Regional and Business Studies*, *12*(2), 33-46.

Darmawansah, D., Hwang, G. J., Chen, M. R. A., & Liang, J. C. (2023). Trends and research foci of robotics-based STEM education: a systematic review from diverse angles based on the technology-based learning model. *International Journal of STEM Education*, *10*(1), 1-24.

Fleming, A. D., Goatman, K. A., Philip, S., Prescott, G. J., Sharp, P. F., & Olson, J. A. (2010). Automated grading for diabetic retinopathy: a large-scale audit using arbitration by clinical experts. *British Journal of Ophthalmology*, *94*(12), 1606-1610.

Hughes, J. (2021). The Deskilling of Teaching and the Case for Intelligent Tutoring Systems. *Journal of Ethics and Emerging Technologies*, *31*(2), 1-16.

Holford, J., Milana, M., Webb, S., Waller, R., Hodge, S., & Knight, E. (2022). Shaping the field of lifelong education through three critical debates in the International Journal of Lifelong Education. *International Journal of Lifelong Education*, *41*(6), 549-571.

Jha, P., & Parvati, P. (2020). National education policy, 2020. *Governance at Banks*, *55*(34), 14.

Jindal, P., & Tej, F. (2016). Being an educator in private vs. public school system: A comparative study on job contentment. *IRA International Journal of Education and Multidisciplinary Studies*, *3*(3), 309-319.

Kumari, S., Jindal, P., & Chand, P. K. (2022). Social Media Networking: As Facilitator of Acculturation. *ECS Transactions*, *107*(1), 9381.

Kim, J. (2022). The Interconnectivity of Heutagogy and Education 4.0 in Higher Online Education. *Canadian Journal of Learning and Technology*, *48*(4), 1-17.

Küfeoğlu, S. (2022). SDG-11: sustainable cities and communities. In *Emerging Technologies: Value Creation for Sustainable Development*, 385-408, Cham: Springer International Publishing.

Küfeoğlu, S. (2022). SDG-12: Responsible Consumption and Production. In *Emerging Technologies: Value Creation for Sustainable Development*, 409-428, Cham: Springer International Publishing.

Lee, S. Y., & Jung, M. (2021). Exploring competing perspectives on how to design open innovation program for high school STEM education: A case study. *Education Sciences*, *11*(7), 322.

Levin, B. A., Piskunov, A. A., Poliakov, V. Y., & Savin, A. V. (2022). Artificial Intelligence in Engineering Education. *Higher Education in Russia*, *31*(7), 79-95.

Ling, L. S., Pang, V., & Lajium, D. (2019). The planning of integrated STEM education based on standards and contextual issues of Sustainable Development Goals (SDG). *Journal of Nusantara Studies (JONUS)*, *4*(1), 300-315.

Marshall, A. G., Vue, Z., Palavicino-Maggio, C. B., Neikirk, K., Beasley, H. K., Garza-Lopez, E., & Hinton Jr, A. (2022). The role of mentoring in promoting diversity equity and inclusion in STEM Education and Research. *Pathogens and Disease*, *80*(1), ftac019.

Maruf, I. R., Nugroho, B. S., Kurniawan, A., Musiafa, Z., & Satria, E. (2022). Virtual Learning Apps: Best Instructional Leadership Practices in the Digital Age Efforts to Improve Student Learning Outcomes. *Jurnal Iqra': Kajian Ilmu Pendidikan*, *7*(1), 32-43.

Nnorom, N. R. (2016). Effect of investigative laboratory approach and expository method on acquisition of science process skills by biology students of different level of scientific literacy. *African Journal of Education, Science and Technology*, *3*(1), 236-243.

Obianuju, O. S., Obiajulu, A. N., & Ella, F. A. (2013). Science education for sustainable development in Nigeria: challenges and prospects. *Academic Journal of interdisciplinary studies*, *2*(6), 159.

Ostreika, A., Zailskaitė-Jakštė, L., Platužienė, J., Rutkauskienė, D., & Punys, V. (2020). Stem education challenges: problem identification from teachers and students' perspectives. In *ALTA'20, 73-81.*

Panigrahi, M. S. P., & Gupta, S. (2021). *National education policy-2020: rethinking assessment in higher education*. 9(46), 11256-11263.

Roy, S., & Ayalon, L. (2022). Intergenerational Relations in the Climate Movement: Bridging the Gap toward a Common Goal. *International Journal of Environmental Research and Public Health*, *20*(1), 233.

Sharma, A., Zhang, Z., & Rai, R. (2021). The interpretive model of manufacturing: a theoretical framework and research agenda for machine learning in manufacturing. *International Journal of Production Research*, *59*(16), 4960-4994.

Shashidharan, M., Bansal, R., Hothi, B. S., Athavale, V. A., Mahajan, Y., & Anwar, S. (2021). A review on National Education Policy 2020 and its influence on academics. *J. Legal Ethical & Regul. Isses*, *24*, 1.

Szczotka, M., & Szewczuk, K. (2021). Students of Teacher Training Programs and Their Views on STEM: A Focus Interview. *Elementary Education in Theory and Practice,* 16, 3(61), 115-137.

Tytler, R. (2020). STEM education for the twenty-first century. *Integrated approaches to STEM education: An international perspective*, 21-43.

Wirzal, M. D. H., & Putra, Z. A. (2022). What is the correlation between chemical engineering and special needs education from the perspective of bibliometric analysis using vosviewer indexed by google scholar. *Indonesian Journal of Community and Special Needs Education*, *2*(2), 103-110.

Zhong, F. (2018). Designing Adaptive Learning Objects for Enhanced Student Engagement in Data Structures and Algorithms. In *Proceedings of the 49th ACM Technical Symposium on Computer Science Education*, 1089-1089.

Zouda, M., El Halwany, S., & Bencze, L. (2023). Science and technology studies informing STEM education: Possibilities and dilemmas. In *Challenges in Science Education: Global Perspectives for the Future*, 201-227, Cham: Springer International Publishing.

Chapter 4

An Assessment and Evaluation of STEM Outcomes in the Education Sector

Deepa Sharma*, PhD
Ram Singh, PhD
and Sunayna Khurana, PhD
Department of Maharishi Markandeshwar Institute of Management,
MMDU Mullana, Ambala, India

Abstract

STEM is an interdisciplinary method of education that combines academic ideas with lessons from the actual world. Students use science, technology, engineering, and math in situations that connect the classroom to the outside world. The Next Generation Science Standards, which were created to enhance how students understand and apply science, highlight the value of STEM in education. In the classroom, project-based learning is often the main focus of STEM education. To stress the application of science and better prepare students for future classes, projects, and activities use technology. Here, we identify significant developments in education that start with a fundamental acceptance of learning goals, including interest, engagement, and identity-building, that go well beyond specific assessments of knowledge and abilities. Within that paradigm, we observe the emergence of common tools and similar constructs across numerous informal STEM education settings. We also highlight developments in stealthy technologies and potent analytical methods that enable more direct evaluation of learners' evolving experiences than ever before. Finally,

* Corresponding Author's Email: deepa.sharma@mmumullana.org.

In: STEM
Editors: Priya Jindal, Amandeep Singh and Rashmi Aggarwal
ISBN: 979-8-89113-305-1

we discuss some of the learning ecosystem metaphors that guide the professional community of informal STEM learning evaluators and the underlying mechanisms that enable its development. Are policy changes needed in the current STEM-related system to find well-thought-out, systematic, and connected policies that address the challenges ahead: improving student outcomes, building the STEM teacher workforce, Rethinking the curriculum, etc.?

Keywords: STEM education sector, performance, learning, outcomes, challenges

Introduction

STEM stands for, Science, Technology, Engineering, and Mathematics. Despite the quick advancement of technology and the creation of new tools, the number of people who actually benefited from these tools was inadequate to support the creation of new ones.

It was anticipated that STEM education would address this issue. Additionally, students' passion for STEM fields has been waning over time. It was crucial to increase students' interest in and approaches to STEM topics and to encourage them to pursue careers in the sector. Only in this way could nations produce people who were ready for future competition, developed in terms of work competencies, keeping up with the times, and producing innovators and entrepreneurs. As a result, many nations have been attempting to incorporate STEM education into their curriculum. Improves students' ability to investigate, solve issues, analyze data and evidence, ask questions, draw conclusions, articulate discoveries, and the effective pedagogical approach of inquiry-based learning. Inquiry-based learning strategies help students engage in authentic, relevant, and relevant real-world learning across all STEM disciplines. The ideas of design-based learning have promoted inquiry-based learning when engineering and technology are integrated into STEM education. Design-based learning emphasizes the creation of novel artifacts, unique solutions, and innovative systems. Students work through reflective thought processes and applications while tackling real-world problems. Making predictions, iterative decision-making, and authentic learning are all topics covered. Future career decisions are influenced by secondary school students' attitudes and interests in science and mathematics. This is the reason why the strong interest of students in STEM subjects acts as a motivator for them to pursue careers in these fields. A key factor in reducing student enthusiasm for careers in STEM disciplines is the lack of initial knowledge about the employability of these fields. To prepare for the STEM workforce of the future, it is important to explore the interests and career objectives of middle school students in STEM subjects. When STEM education first appeared, it attracted the interest of many organizations. Botanists were thrilled because they believed that teachers had finally understood the importance of a major component of plants. Technologists and engineers were thrilled as they believed it was a reference to a clock component (Jindal et al., 2016). Indeed, none of these ideas about STEM were based on what is currently being used for science, technology, engineering, and math education. STEM is usually taken to refer to science or mathematics.

This is rarely discussed in engineering or technology, and it is a problem that has to be fixed. Educators in the STEM society's determination need to explain what the contraction earns for learning policies, initiatives, and practices if STEM education is to become more than just a phrase. The discussion that follows outlines some of the potential implications of STEM for modern schooling. The discussion that follows outlines some of the potential implications of STEM for modern schooling and the relevance of science technology and engineering in educational programs. The goal of STEM education is to help students develop strong problem-solving abilities, innovation, confidence, logical thinking, and STEM literacy. STEM literacy is evolving from science literacy. The use of science, technology, engineering, and math in education is made possible by STEM literacy, an interdisciplinary educational strategy that helps people become more competitive in their chosen fields of work. The definition, application, and integration of STEM concepts are essential to tackling difficult problems and repurposing them to find solutions. STEM literacy is important for students, as a first step toward their future careers. It is at the core of the essential fundamental skills. Specialized presentation of STEM components throughout the high school and university years supports the development of students' knowledge in science, technology, engineering, and math fields. An instructional paradigm that can support students' development of STEM literacy. Students are required to apply their knowledge to real-world situations using engineering design concepts, and they do so using project-based STEM elements. Along with engineering design capabilities, analytical thinking, and problem-solving abilities will be strengthened. Additionally, it will meet the standards set by politicians and encourage students to pursue STEM-related careers. Therefore, the focus should be on STEM literacy. The ability to use STEM principles to address issues is known as STEM literacy. According to several studies, students' attitudes, interests, and motivation in STEM can be increased through extracurricular STEM activities. However, most of these activities are transient and loosely planned. It is recognized that a structured, inclusive STEM curriculum is essential for students in preschool, elementary school, and middle school. Incorporating STEM into general lessons and curriculum increases STEM literacy. Students can follow jobs, interests, and perspectives in STEM subjects with the help of a structured curriculum. STEM education should emphasize social values, be applicable, be relevant to the economy, be relevant to national education, and be scientific.

To successfully integrate STEM learning experiences in their classrooms, teacher applicants and experienced educators should possess subject-specific knowledge in addition to teaching abilities and confidence across all subject areas. It is evident that teachers need ongoing training to increase their preparation for teaching STEM subjects. At the university level, there isn't any course material related to STEM education. Outside-of-class activities help teachers engage students in STEM. The teachers who are responsible for implementing STEM programs do not have enough STEM understanding. Many primary school instructors lack expertise and confidence, as well as have inadequate plans for teaching STEM to students. It would be suitable to enroll in a STEM learning course at the basic and middle school levels because of this. There is a need for a strong, integrated policy that involves school education, even though education cannot shoulder the full burden and cost of growing STEM capacity on behalf of the country's enterprises and industries. The student results in STEM, the STEM teacher workforce, and the STEM curriculum are the three key areas where the school sector has the best possibility of making a difference. The purpose of this study was to learn more about the attitudes, background knowledge, teaching strategies, and inherent difficulties in STEM education that Chinese teachers of STEM-related disciplines have. Demographic information about the participants, such as gender, seniority, grade, and principal subjects taught, as well as their background in STEM education and training, was also gathered. China currently lacks a formal training system and a structured admission process for the certification of new STEM instructors. STEM educators are typically focused specialists who cover degrees in fields ranging from information technology to mathematics and science-related courses. They have worked on STEM-related projects, initiatives, or programs including running STEM competitions, STEM tutoring, or STEM extracurricular lessons on campus. Teaching students in STEM-related competitions or presenting STEM lessons as on-campus extracurricular activities and included in STEM training courses designed primarily to prepare and instruct subject teachers in the implementation of STEM education in the classroom May or may not be. Finally, STEM educators' thinking, awareness, support, and instructional strategies reflect the challenges teachers express when teaching STEM.

Literature Review

Making is widely recognized in STEM education as a kind of hands-on and active learning. To foster students' development as creative STEM thinkers, several schools have adopted making as a teaching approach with a heavy emphasis on designing, doing, and creating (Ernst and Glennie, 2015).

The ability to come up with several answers to a problem through divergent thinking rather than just one through convergent thinking is referred to as creativity (Karakaya and Yılmaz, 2022). With the help of maker pedagogy, students might be able to investigate a topic as a team, share ideas among themselves, and develop and rebuild their concepts. Additionally, creativity serves as a crucial instrument that encourages students to implement their fresh insights and advance their creative competency through innovation ((Maass et al., 2019). STEM maker projects can effectively foster kids' creativity in a classroom setting. Students can use 3D printing to create imaginatively created dolls by working in groups and sharing their ideas with classmates (Becker and Park, 2011). Similarly, when given the chance to explore various solutions and get suggestions from others, high school students participating in a workshop on circuit design came up with new ideas for their artifacts of electronic textiles (Martín-Páez et al., 2019). In general, maker pedagogy encourages a creative environment and helps students see creative new possibilities. Experiential learning is integrally connected to pedagogy, which includes hands-on activities and inquiry-based learning (Peralta et al., 2013). Based on the constructivist paradigm, which holds that learning occurs when students experience something and reflect on it, this form of education (Bybee, 2010).

Teachers provide students with real-world situations, and students then identify issues, devise potential solutions, and suggest evaluation techniques concerning the difficulties (Carter, 2017). To approach problems with creativity and imagination, the students engage in in-depth research using materials, objects, and concepts as well as engaging ways to derive meaning and knowledge from their experiences. To approach problems with creativity and imagination, the students engage in in-depth research using materials, objects, and concepts as well as engaging ways to derive meaning and knowledge from their experiences. Observations and reflections are a part of the overall problem-solving process. They must be able to engage in sophisticated activities because they must gather and coordinate their own, other people's, and material resources to meaningfully participate in teaching (MacDonald et al., 2020). Young children must therefore have access to and a developing relationship with teacher/mentor help and expertise to make learning a process. To form an opinion, one must be able to think clearly and logically. Students must evaluate solutions, link ideas from several fields, and understand complicated situations (Strimel and Grubbs, 2016). The real world frequently sees seamless collaboration among STEM fields. The connections between disciplinary subjects and everyday life provide learning opportunities with local and international populations, demonstrate how learning is applied in real-world situations, and involve students in meaningful, authentic learning. These lead to increased cognitive engagement, which supports their capacity for critical thought. These lead to increased cognitive engagement, which supports the development of their critical thinking.

Subramaniam et al., (2012) found that 645 secondary school students worked on real-world issues that honed their critical thinking skills; and found that students' exposure to real-world issues enhanced the development of their critical thinking skills. It might be because real-world issues provide students with cognitive challenges and need they to carefully analyze all of their options for solutions as well as how those solutions fit into societal contexts (Morrison et al., 2021). Students must work creatively and critically in groups to propose potential answers to these difficulties. Two important indicators of students' higher education and career plans in STEM-related disciplines are STEM passion and identity (Gomez and Albrecht, 2013). They allow people to acquire personal significance from their efforts in the STEM sector and reflect individuals' self-images. Therefore, two key learning outcomes of maker pedagogy in STEM classrooms are increasing students' interest in STEM and identity.

STEM teachers have ideas about the responsibilities and duties of teachers and students, how students learn science, and how teachers may improve science instruction (Allen and Peterman, 2019). The second approach is concerned with conceptions of the environment of knowledge. These conceptions can be broadly categorized into ontological and epistemological beliefs. According to them, the former refers to ideas on the reality or existence of scientific matter, whereas the latter deals with what knowledge is and how it is created, acquired, and acceptable. A broad focus is also included in the definition of the knowledge base of STEM teachers. Three types of teacher knowledge were outlined by (Allen and Peterman, 2019) propositional knowledge, case knowledge, and strategic knowledge. A declaration of the fundamental ideas guiding teaching and learning contains propositional knowledge.

STEM-Related Challenges Faced by Educators and Students in the Education Sector

"STEM teachers' internal challenges" refers to a situation that makes it difficult for teachers to pursue professional development and reach their goals. It is linked to the personal competence of instructors, instructional methods, academic difficulties, curriculum difficulties, student concerns, evaluation, and teachers' time and expertise. The education sector has assessed and described STEM literacy challenges based on a set of policies and programs, classified as follows in Table 1.

Table 1. STEM-related challenges faced by educators and students in the education sector

Student	Teachers	Curriculum
a. Senior STEM subject choices b. Concentrate on senior secondary performance c. Selecting a career d. Equity programs that target racial and gender representation	a. The competence of science lecturers and instructors b. STEM-trained instructors c. Instructors' ongoing professional development d. Network and Center for National STEM Education	a. Curriculum development for STEM b. A huge diversity of curriculum enrichment coordination

Source: Author's compilation.

The application of teaching methods and the use of instructional materials present obstacles for STEM teachers. The three primary difficulties educators encounter when introducing STEM techniques in their classrooms are as follows: A lack of understanding of STEM, a challenge linking scientific and mathematical concepts, and a problem with the STEM approach's applicability to all topics. Based on teachers' attitudes and knowledge about classroom practices in connection to the inherent obstacles they face, the theoretical outline developed in this study focuses on the instructional accomplishment of STEM. The following points are discussed below:

1. A lack of awareness of the multidisciplinary character of STEM-based curriculum, particularly a lack of understanding of how to successfully incorporate STEM-related subject matter areas:
2. Be deficient in comprehension of the standards and content in topics outside of their majors, particularly a misunderstanding of what engineering education entails;
3. Lack of time, particularly for mutual arrangement, integrating information starting many disciplines, creating STEM core curriculum activities with other teachers, and putting STEM philosophy and culture behavior into practice;
4. The impact of organizational and structural factors in schools; the results of conventional exams;

Educators' opinions about the trans-disciplinary nature of STEM should potentially tie classroom instruction to real-world situations, enhance students' technical literacy, and develop their aptitude for STEM-related occupations. How to evaluate the information received from teachers about the many difficulties associated with STEM teaching and how they should apply ideas about the underlying difficulties and impact on instructional skills to explain scientific and technological concerns related to STEM, in addition to the foundation of scientific knowledge Needs to be held firmly.

Structure of STEM Education

Best practices in science, technology, engineering, and math (STEM) education are to be found in the STEM Education Framework. It incorporates cutting-edge and successful methods used all around the world.

The Structure of STEM Education Is Based on 3 Elements

a. Core Competencies: How many opportunities are given to pupils to acquire the 21st-century abilities necessary to succeed in the contemporary workplace? The key points of Core Competencies:

1. Students should be given the chance to assess a variety of information, evidence, and primary sources and choose appropriate evidence to support claims; evaluate the work of others; and differentiate between fact and opinion.
2. Students' hypotheses should be tested through activities that apply the scientific method.

When formulating arguments, critiques, or hypotheses, students are encouraged and given the opportunity to reason about issues, experimental techniques, and phenomena using one or more approaches such as computational, systems, or design thinking are being given. Supporting materials are used to encourage teacher-student dialogue and argument.

3. Although they are not essential elements of instructional activities, students have the chance to assess information, evidence, and original sources; and utilize evidence to support an argument.
4. Students have the chance to draw conclusions or generalizations about phenomena and make predictions based on the information provided.
5. Students are encouraged and given the chance to use one or more perspectives such as computational, systems, or design thinking to analyze issues and occurrences.
6. Students should be given a chance to assess the facts or information presented.

b. Instructional Design: To what extent do the program's design and/or the materials' content reflect research-based pedagogy and a well-coordinated set of learning goals, resources for supporting learning, and tools for assessment?

By ensuring that every teacher has the chance to understand fundamental STEM concepts, such as computational thinking, and to become digitally literate, we can provide a stronger foundation for STEM literacy (Kumari et al., 2022). People who are knowledgeable in STEM fields will be better able to cope with the rapid pace of technological change and will be more prepared to participate in civil society. Promote diversity, equity, and inclusion in STEM occupations, while providing access to lifelong STEM education for all Americans, especially those who have traditionally been underrepresented in STEM fields and occupations. The STEM workforce of the future can be prepared by developing authentic learning experiences.

c. Implementation: To what extent are the supports or services required for distribution and successful implementation made available? STEM education is integrated into the harmonious knowledge base and skill set needed for the 21st-century economy. The best STEM education uses an interdisciplinary method of teaching, where complex academic ideas are combined with practical applications and students apply STEM in situations that connect their learning to their community, workplace, and the larger world. Leaders in STEM education are continually broadening and deepening their focus and expanding subjects of study beyond just a mix of the four sciences to include the humanities and arts. Modern STEM education imparts practical qualities such as tenacity, resilience, higher-order thinking, and estimation, in addition to abilities such as critical thinking, problem-solving, and design.

Anyone involved in STEM education curriculum creators; content creators, instructors, students, parents, school administrators, policymakers, and philanthropists can use this framework to help guide high-quality educational creation and assessment.

Need and Support Felt by Educators to Apply STEM

Some research documented the potential areas where teachers would want extra assistance. The research that addressed this need for support identified five key categories. Collaboration, curriculum, district support, prior experiences, and professional development were among them.

Collaboration

Teachers believe that STEM programs would be more successful if there was a culture of collaboration. Teachers stressed the value of working together with other STEM educators and academic experts to not only foster an environment that facilitates preparation for STEM lectures but also to provide students with a teamwork model. STEM pedagogy demands that students work together to solve problems, so it is advantageous for a teacher to demonstrate the power of a team effort. One educator stated that due to historical teacher silos, STEM planning and implementation need an integrated team approach. For successful transdisciplinary teaching, which is essential for STEM subjects, teachers believe it is based on teamwork. Collaboration between subject-area instructors can be facilitated through technology-enabled networks. Giving teachers the space, time, and opportunities for open dialogue and collaborative preparation can be critical to the success of implementation.

Syllabus

Educators feel that having access to high-quality curricula will increase the likelihood of STEM projects being successful. Educators talk about the value of an engineering-based, adaptable curriculum. The curriculum should be adaptable enough to be used in a variety of educational settings and with students of various skill levels while focusing on the engineering design process. Teachers felt that such a curriculum improved their self-efficacy, or ability to teach STEM. Additionally, there should be a clear and direct link between the STEM curriculum or module and the standards. Additionally, the development of these modules should be considered. Teachers reported a need for special, pre-made STEM challenges they could use in their classrooms right away. These issues need to be rooted in STEM fields. Additionally, the curriculum needs to be conscientiously implemented so that teachers use the standards and objectives that the curriculum makers intended.

Past Experiences

Teachers believe that prior experience with student-centered, inquiry-based approaches to instruction makes STEM initiatives more successful. Teachers' earlier, similarly organized experiences were frequently viewed as catalysts for STEM success. Teachers who had taken more science or math courses in college or who had used similar teaching strategies such as guided independent research studies, problem-based learning, inquiry-based learning, and questioning techniques felt that these experiences had given them the ability to encourage the inductive and deductive reasoning across disciplines that are required for STEM. These previous experiences led to an improvement in confidence with STEM pedagogy.

Professional Growth

Teachers think that successful STEM efforts would be facilitated by well-planned and frequent professional learning opportunities. Learning opportunities for teachers to improve their capacity to successfully incorporate STEM information into their curricula were mentioned as the support that would increase the efficacy of STEM education most frequently. After taking part in professional development programs, teachers at all career levels reported significant gains in their efficacy, knowledge, and confidence to teach STEM. The goal of effective professional development or continuing education is to improve teachers' content knowledge and STEM experiences while giving them the time and structure to investigate how STEM may be incorporated into their curriculum. According to research, these elements will directly affect how teachers teach and how students learn. Teachers wished there were additional methods for enhancing students' success in engineering design tasks. When integrating engineering into math and science lessons, teachers require assistance. Teachers believed that continuing support for planning and implementation should include pedagogical skills they can utilize to improve academic success for students.

Role of STEM Outcomes in the Education Sector

Everyone familiar with the idea of STEM education, the main goals and objectives of STEM are explained below:

1. Access to Knowledge and Insight: STEM programs are designed to help students access information and insight related to science, technology, engineering, and mathematics. IDEA enables students to investigate a variety of topics and projects.
2. Implementation of Best Practices: Students should be able to implement best practices in all STEM fields. The long-term objective is to improve one's competency in these four areas.

3. Knowledge Acquisition and Creativity: Students must become knowledgeable in four STEM disciplines and creative thinkers.
4. Meeting Demands in the Global Workplace: STEM education is not just accepted in India. It is widely regarded. A graduate who used a STEM approach earns acclaim on a global scale. Since they can handle the demands of the workplace, they can apply for jobs anywhere in the world.
5. Advantage for Students: Students who study STEM have an advantage over those who have learned through the standard educational system since STEM concentrates on four specific areas. The likelihood that a student majoring in STEM will be employed at a substantially better salary because of the gained skill set is maybe the most crucial factor.
6. Inventions in the STEM Fields: The STEM fields allow students to learn both theoretically and practically. Students can achieve "Invention," the greatest level of the educational plan, by combining the two components. STEM enables pupils to cultivate creative ability.
7. Leadership Qualities: Over time, students who have adopted the STEM method develop the efficiency necessary to address any situation. They view obstacles as chances to advance both personally and professionally.

The Foundation of STEM in the Education Sector

1. Solution-based Learning: The students in this style of learning gather data, analyze it, develop scientific hypotheses, and pinpoint issues to carry out research tasks. They also create a prototype to address the issue before coming up with the best answer. Students can then use technology and mathematics to solve issues in the real world.
2. Systematic Education: To encourage pupils to challenge their preconceptions and engage in critical thought, rigorous learning must be applied to the teachings. It prevents people from studying memories and recollecting knowledge for tests.
3. Life skills, Career, and Technology: Making the kids proficient in the subjects they are learning is part of this. It places a strong emphasis on project-based learning, internships, and practical experiences for

students. The student's critical thinking, teamwork, communication, and creativity are all goals of the program.
4. Personalization of Learning: Personalized learning means switching from teacher-led classrooms to more student-centered learning environments. In this, the student has the option to select their subjects, learn at their own pace, and take tests as needed.
5. School community and Belonging: Students who engage in this form of learning learn about inclusiveness, respect, and accepting others for who they are. It seeks to provide students with the chance to talk about their thoughts and experiences with other students and learn from their viewpoints.
6. Connection to the Larger and External Community: This entails encouraging pupils to be inspirational and global thinkers. Local environments and community groups are made available to the students as learning tools. The children are aware of problems in their community and work to resolve them.

Reasons for a STEM-Based Course

Many traditional public and private schools teach a wide range of subjects, including science, social studies, history, English, language arts, mathematics, fine arts, economics, and physical education. For students to gain any sort of competitive advantage in the modern world, all they need is a new educational formula. Some reasons are described below:

1. There will be greater demand for STEM-related careers

STEM-related occupations are in high demand. The world requires a strong group of engineers, technicians, mathematicians, and scientists. The number of people who can support the medical world is increasingly outpacing it. The popularity of electronics, computers, televisions, audio, and other devices is soaring, yet there aren't as many people interested in working in these fields. If more people do not step up and pursue these vocations, the existing gap could make the United States weak in these areas.

2. How students react to teachers

The chance for teachers to showcase their actual talents and passions must be provided. It is a well-established truth that pupils are influenced by their professors' daily behaviors and emotions. Leaders who inspire others are happy teachers. Teachers must be given the latitude, resources, and guidance they require to illustrate why each of those areas is fascinating to make STEM-related occupations seem enticing.

3. People who work in STEM careers earn more money

These careers are paying more money to lure students to study them due to the severe lack of STEM-focused students. This is logical. Financial security motivates people, STEM-related occupations are expanding, and companies will go to any lengths to fill a job opening. Everyone in this circumstance is positioned for success.

4. There is a growing need for technical expertise

This implies that to be employable, all students regardless of the vocational concentration they decide on will need STEM-based studies. Implementing a STEM-based education gives all students a competitive advantage in the modern workplace, not only those who want to pursue jobs in STEM fields.

5. Anyone can enter a career in the STEM field

A doctoral degree in one or more of the listed fields is not necessary for all STEM-related employment. Many careers will hire someone with just a high school diploma, a master's degree, a bachelor's degree, an associate's degree, or an associate's in applied science. For students who cannot immediately afford a degree, many STEM-related occupations offer on-the-job training to get them employed and financially stable enough to build a life for themselves.

6. There are scholarships for students with a STEM focus

Any student with a strong will to succeed will be successful in anything he/she sets his/her mind to. Students with a focus on STEM-related fields of study are given special scholarship opportunities to support their distinctive and important intellectual achievements.

Facts Behind the STEM Education

The cornerstone of the most useful abilities in the current Information Age is thought to be STEM education. The abilities acquired through this academic approach are very useful in everyday life. Therefore, a child's holistic development depends on a solid foundation in STEM stands for Science, Technology, Engineering, and Math topics. When educationalists understood how crucial it was for students to obtain technical abilities to keep up with the rapid advancement of information technology, the STEM learning approach was formally launched in 2001. Since then, it has evolved into a crucial component of our curriculum. Some vital facts are discussed below:

1. Helps students to crack the highest paying jobs

Innovative and critical-minded individuals are welcomed in the highest-paying job markets. STEM fields are better prospects in terms of pay scales as compared to non-STEM occupations. Here's a look at some of the top STEM career options and their median salary along with the job growth expected in the upcoming years.

2. STEM fields provide female students with a sense of empowerment

The future of women's and girls' emancipation lies in the STEM disciplines. Since these are the options with the biggest salaries. In this Information Age, Unfortunately, there is still a long way to go before gender equality in these professions. Few female students attend higher education institutions to study STEM fields. Hopefully, in the years to come, this gap will be closed. And the empowerment of STEM education will enable our female students to succeed in the employment markets. To counter the pandemic's evil effects, the post-Covid employment landscape currently needs female authority. Because overcoming the greatest economic recession in history is currently a struggle. The future of the world is undoubtedly in STEM vocations. And they are excellent for increasing the proportion of female workers in the workforce.

3. COVID-19 has significantly increased the number of STEM jobs

According to the Bureau of Labor Statistics, COVID-19 has increased interest in STEM careers. With a significant increase in information technology and computer-related jobs. It draws attention to changes in the labor market at this time, with increasing demand in sectors including software publishing, data processing, hosting, and related services, computer systems design and related services, and scientific research and development services. STEM education gives kids exposure to the world, critical thinking skills, and an inventive mindset that enables them to contribute significantly to society.

4. STEM experts successfully solve global concerns

The world could potentially be changed by STEM education. Because it fosters the ideals, knowledge, and curiosity that are currently in demand. Students who study STEM develop advanced levels of comprehension. Where they develop a certain amount of cognitive flexibility and work for the welfare of society as a whole. The emotional intelligence of students who enroll in these courses is higher. This implies that they are more aware of the world. They typically have highly developed soft skills that can be transferred, such as problem-solving, critical thinking, advanced communication, adaptability, time management, etc., that make them ideal leaders and autonomous workers in any situation. These society members can communicate with people from different cultural backgrounds. And in doing so, they offer their services to the world community when it needs them.

Current Scenario of STEM in Today's Era

1. STEM encourages an integrated study style that focuses on learning themes rather than courses: It integrates project-based, multidisciplinary approaches, with the natural sciences, technology, engineering, and artistic fields serving as the foundation. To eliminate the teaching of the aforementioned subjects independently and abstractly, relying primarily on written sources, STEM education offers a complete makeover of the curriculum. Because these subjects are practically intertwined, it is crucial to teach science, technology, engineering, and math together.
2. STEM education is based on a hands-on method of learning: It instructs children in the actual application of scientific and technological information and provides examples of this application through exercises. Each session teaches students how to create, build, and market original goods. They learn through working on a

particular project, and as a result, they make a physical prototype of a real product.

3. STEM education develops critical thinking and problem-solving abilities, which are crucial for helping kids deal with challenges they may encounter in life.
4. Fostering a child's self-confidence is one of the key objectives of STEM education: The actual application of this concept in STEM education gives children a sense of moral accomplishment and boosts their self-esteem.
5. Teamwork is encouraged through STEM education: It enables them to collaborate, freely express their views, have discussions, support their positions, and come to certain decisions as a group.
6. STEM education spurs a surge in interest in technological fields worldwide: Students will learn that pursuing careers in engineering or mathematics is not a ticket to a life of menial labor but rather an exciting and enjoyable process that can have a positive impact on one's future.
7. Visualization is emphasized in STEM education: Visualization tools, such as posters, whiteboards, flipcharts, *etc.,* are frequently utilized in the classroom to aid students in better understanding difficult technical and scientific subjects. Students benefit from this in that they can express their thoughts and proposals while also honing vital management and presentational skills.
8. STEM training fosters a specific way of thinking: The child's learning becomes more constant, but he or she is also not scared to consider other, unconventional options. In the classroom, kids actively seek out solutions to difficulties. This develops the capacity to organize tasks effectively and break them down into manageable pieces. Additionally, kids learn how to evaluate information, take part in fruitful debates, and support their opinions.
9. STEM education gives kids considerably more independence than traditional schooling does: The learning process is unaffected by relationships with teachers or peers; progress can then be evaluated more objectively. STEM teaches a youngster to take more ownership of their education and to not rely on others for their achievement.
10. STEM allows for the individualization of learning: The majority of top colleges in the world tailor their curricula so that students only have to take a few foundational subjects and are free to choose the

rest. Each student's potential is shown through this method, which also enables them to better meet the demands of possible employers.

Conclusion

The main conclusions of this study suggest that the inherent difficulties experienced by educators can be predicted by students' background familiarity and thinking, which are mediated by STEM teaching strategies. Compared to a teacher who only uses their primary discipline, a teacher who comprehends the environment, explanation components, and pedagogy of STEM education is more likely to experience lower-level essential obstacles of STEM educators. The analysis demonstrates a substantial correlation between beliefs and practices and knowledge, but not as strong a correlation between beliefs and perceived inherent problems. The actuality of integrating the STEM syllabus seems to at hand a barrier to instructors, which is understandable given that teachers' ideas and understanding control how educators educate STEM. The instructional techniques that teachers use are greatly influenced by their beliefs and worldviews; it is clear from this study that beliefs have a significant impact on STEM instructional practices. Innovative approaches to teacher training and professional development are required to close the gap between beliefs and the inherent difficulties that teachers experience. The evaluation of non-formal STEM education is the research topic addressed by this study. The findings concluded that assessment was not given the same priority as assessment in non-formal STEM education. To provide systematic and effective non-formal STEM education, assessment and evaluation must be able to accurately reflect non-formal STEM education. Assessment and evaluation of non-formal education can be done together, not just one of them. This is done to show not only how valuable and worthwhile it is, but also how far students have progressed in their learning, as shown through the assessment process another finding is that STEM education that is not formal is valued. On the other hand, this result of this learning also raises the possibility of an association connecting STEM teachers' perceptions of intrinsic difficulty and their training in STEM fields. With the incorporation of STEM disciplines and adequate resources for such pedagogical approaches, teachers are required to carry the STEM educators and knowledge environment. The dynamics of teachers transitioning to cutting-edge pedagogical techniques and successfully integrating the many disciplines should be supported through professional development. Teacher professional development frequently supports self-efficacy and lessens difficulties related to STEM skills. Making innovative decisions and navigating the intricacies of teacher professional development to attain the ideal STEM focus is essential. Overall, based on the results of this study, Chinese STEM teachers' integration of STEM disciplines is currently facing several challenges. This study also raises the possibility of a link between STEM teachers' perceptions of intrinsic difficulty and their training in STEM fields. With the amalgamation of STEM disciplines and adequate wealth for such instructive methods, teachers are required to maintain STEM educator and education. The dynamics of teachers transitioning to cutting-edge pedagogical techniques and successfully integrating the many disciplines should be supported through professional development. Teacher professional development frequently supports self-efficacy and lessens difficulties related to STEM skills. Making innovative decisions and navigating the intricacies of teacher professional development to attain the ideal STEM focus is essential. Overall, based on the results of this study, Chinese STEM teachers' integration of STEM disciplines is currently facing several challenges.

References

Allen, S., & Peterman, K. (2019). Evaluating informal STEM education: Issues and challenges in context. *New Directions for Evaluation*, *2019*(161), 17-33.

Bybee, R. W. (2010). Advancing STEM education: A 2020 vision. *Technology and engineering teacher*, *70*(1), 30.

Becker, K. H., & Park, K. (2011). Integrative approaches among science, technology, engineering, and mathematics (STEM) subjects on students' learning: A meta-analysis. *Journal of STEM education: Innovations and research*, *12*(5).

Carter, L. (2017). Neoliberalism and STEM education: Some Australian policy discourse. *Canadian Journal of Science, Mathematics and Technology Education*, *17*, 247-257.

Ernst, J., & Glennie, E. (2015). Redesigned high schools for transformed STEM learning: Performance assessment pilot outcome. *Journal of STEM Education*, *16*(4).

Gomez, A., & Albrecht, B. (2013). True STEM education. *Technology and Engineering Teacher*, *73*(4), 8.

Jindal, P., & Tej, F. (2016). Being an educator in private vs. public school system: A comparative study on job contentment. *IRA International Journal of Education and Multidisciplinary Studies*, *3*(3), 309-319.

Kumari, S., Jindal, P., & Chand, P. K. (2022). Social Media Networking: As Facilitator of Acculturation. *ECS Transactions*, *107*(1), 9381.

Karakaya, F., & Yılmaz, M. (2022). Teachers' views on assessment and evaluation methods in STEM education: A science course example. *Journal of Pedagogical Research*, *6*(2), 61-71.

Martín-Páez, T., Aguilera, D., Perales-Palacios, F. J., & Vílchez-González, J. M. (2019). What are we talking about when we talk about STEM education? A review of literature. *Science Education*, *103*(4), 799-822.

MacDonald, A., Huser, C., Sikder, S., & Danaia, L. (2020). Effective early childhood STEM education: Findings from the Little Scientists evaluation. *Early Childhood Education Journal*, *48*(3), 353-363.

Morrison, J., Frost, J., Gotch, C., McDuffie, A. R., Austin, B., & French, B. (2021). Teachers' role in students' learning at a project-based STEM high school: Implications for teacher education. *International Journal of Science and Mathematics Education*, *19*, 1103-1123.

Maass, K., Geiger, V., Ariza, M. R., & Goos, M. (2019). The role of mathematics in interdisciplinary STEM education. *Zdm*, *51*, 869-884.

Peralta, C., Caspary, M., & Boothe, D. (2013). Success factors impacting Latina/o persistence in higher education leading to STEM opportunities. *Cultural Studies of Science Education*, *8*, 905-918.

Subramaniam, M. M., Ahn, J., Fleischmann, K. R., & Druin, A. (2012). Reimagining the role of school libraries in STEM education: Creating hybrid spaces for exploration. *The Library Quarterly*, *82*(2), 161-182.

Strimel, G., & Grubbs, M. E. (2016). Positioning technology and engineering education as a key force in STEM education. *Journal of Technology Education*, *27*(2), 21-36.

Chapter 5

Analyzing Meta-Cognitive and Critical Thinking to Develop Skills and Re-Skills for STEM Literacy

Gagandeep*, PhD
and Jyoti Verma, PhD
Chitkara Business School, Chitkara University, Punjab, India

Abstract

Science, Technology, Engineering, and Mathematics (STEM) is a multi-disciplinary approach that integrates knowledge and skills from different fields to solve complex problems and drive innovation. STEM education aims to prepare students to be creative, critical thinkers, and problem solvers in a rapidly changing world. The present study speaks directly to the role of Metacognitive & critical thinking concerning STEM, rather than the discussion of research within the individual content areas. The study focuses on the skills and reskills development through Metacognitive and critical thinking as seen by academics working in science, technology, engineering, and mathematics and provides concise, relevant research outcomes to improve the quality of integrated STEM education. The study is limited to the above variables in scope and objectives.

Keywords: metacognitive, critical thinking, skills and re-skills development, STEM education

---

* Corresponding Author's Email: gagandeep.1238@chitkara.edu.in.

In: STEM
Editors: Priya Jindal, Amandeep Singh and Rashmi Aggarwal
ISBN: 979-8-89113-305-1

Introduction

STEM education is a real-world, problem-based education that draws on skills and knowledge of fields in science, technology, engineering, and mathematics. Instead, each of these separate disciplines to an equal extent focused on STEM researchers' perspective on how their work contributes to the practical education of STEM via building knowledge, engagement, and innovative skills (MacDonald and Murphy, 2020). STEM sets an important context in which present-day innovation urgently needs to give more diversity and knowledge inclusion (Ahmed et al., 2022; Penprase, 2020). The effective practice in integrated STEM education and contribution to theoretical and practical knowledge gained from diverse approaches focuses on various problems seen by academics working in various fields of Science, Technology, Engineering, and Mathematics (Anderson and Li, 2020).

Here are some ways in which STEM is a multi-disciplinary approach:

1. *Science:* Science is the foundation of STEM education. It provides knowledge and understanding of the natural world, essential for innovation and problem-solving. Science education in STEM integrates principles of biology, physics, chemistry, and other natural sciences.
2. *Technology:* Technology is a critical component of STEM education. It involves using tools, techniques, and systems to solve problems and drive innovation. Technology education in STEM integrates principles of computer science, software engineering, hardware engineering, and other related fields.
3. *Engineering:* Engineering is the application of scientific & technological principles to build innovative solutions to real-world problems. Engineering education in STEM integrates principles of mechanical, electrical, civil, and other engineering fields.
4. *Mathematics:* Mathematics is the language of STEM. It provides the tools and techniques for analyzing data, modeling complex systems, and making predictions. Mathematics education in STEM integrates principles of algebra, geometry, calculus, and other related fields.

STEM is a multi-disciplinary approach that integrates different pedagogies to encourage innovation and problem-solving.

STEM Integrates Pedagogies That Promote Innovation

1. *Inquiry-based learning:* STEM education often emphasizes inquiry-based learning, which involves asking questions, designing experiments, and drawing conclusions based on evidence. This pedagogy encourages students to think critically and creatively and to take an active role in their learning (Johnson et al., 2020).
2. *Problem-based learning:* STEM education also emphasizes problem-based learning, which involves identifying real-world problems and working collaboratively to find solutions. This pedagogy encourages students to apply their knowledge and skills in practical contexts and to develop communication and teamwork skills.
3. *Project-based learning:* STEM education often includes project-based learning, which involves working on long-term, interdisciplinary projects that require students to integrate knowledge and skills from different fields. This pedagogy encourages students to take ownership of their learning, to work collaboratively, and to develop problem-solving and project management skills.
4. *Design thinking:* STEM education often incorporates design thinking, which involves a human-centered approach to problem-solving. This pedagogy emphasizes empathy, creativity, and iterative problem-solving, and encourages students to think outside the box and develop innovative solutions to complex problems.

Role of Meta-Cognitive and Critical Thinking Skills

Meta-cognitive and critical thinking skills are two important cognitive abilities that are crucial in our learning, problem-solving, and decision-making processes. In this chapter, we will discuss these two skills in detail and explore their importance in various aspects of life.

Meta-Cognitive Skills

Metacognition is a tool that helps the student unlock the brain's incredible power and control of learning (Wilson and Conyers, 2016). Meta-cognition refers to the ability to think about one's thinking. It is the process of self-reflection, self-evaluation, and self-regulation of our thoughts, beliefs, and attitudes. Meta-cognitive skills involve monitoring our own mental processes and understanding how we learn, process information, and solve problems. They include:

1. *Awareness of one's learning:* This involves understanding our strengths and weaknesses in learning, identifying the best learning strategies, and evaluating the effectiveness of our learning.
2. *Planning:* This involves setting goals, creating strategies, and organizing the steps necessary to achieve these goals.
3. *Monitoring:* This involves tracking our own progress, evaluating the effectiveness of our strategies, and making necessary adjustments.
4. *Evaluating:* This involves critically analyzing our own performance, identifying areas for improvement, and making necessary changes to enhance our performance.

Importance of Meta-Cognitive Skills

Meta-cognitive skills are vital because they enable us to take control our learning, make informed decisions, and become self-directed learners (Chen, 2022). They also help us to:

1. Improve our memory and recall: By understanding our own learning processes, we can use the most effective strategies to remember and recall information.
2. Enhance our problem-solving skills: By monitoring our own thinking, we can identify potential obstacles and develop strategies to overcome them.
3. Increase our motivation and engagement: By setting goals and monitoring our progress, we can increase our motivation and engagement in learning.

Critical Thinking Skill

Critical thinking is analyzing, evaluating, and synthesizing information to form well-reasoned judgments and make informed decisions (Bradley and Price, 2021). Critical thinking skills include:

1. *Analysis:* This involves breaking down complex information into its parts and understanding the relationship between them.
2. *Evaluation:* This involves assessing the reliability and credibility of information and arguments.
3. *Synthesis:* This involves combining different pieces of information to create a coherent and well-supported argument.

Importance of Critical Thinking Skills

Critical thinking skills are essential because they enable us to make informed decisions, solve problems, and evaluate the credibility of information. They also help us to:

1. *Enhance our communication skills:* By synthesizing information and creating well-supported arguments, we can effectively communicate our ideas and perspectives (Kumari et al., 2022).
2. *Improve our problem-solving skills:* We can develop effective strategies to solve complex problems by analyzing and evaluating information.
3. *Increase our creativity:* By thinking critically, we can generate new ideas and perspectives that may not have been considered.

Relation between Meta-Cognitive and Critical Thinking Skills

Meta-cognitive and critical thinking skills are closely related and often work in tandem to enhance our learning, problem-solving, and decision-making abilities. Meta-cognition reflects one's own thinking processes, while critical thinking involves analyzing and evaluating information to make informed judgments and decisions (Zhou and Wang, 2020). When we use meta-cognitive skills, we can monitor our own thinking and identify potential biases, assumptions, or gaps in our knowledge. This self-awareness can then inform our critical thinking processes, as we can use this information to evaluate the credibility and reliability of information and arguments.

Similarly, we can evaluate information and arguments more objectively and rationally when we use critical thinking skills. This can then inform our meta-cognitive processes, as we can reflect on how we arrived at our conclusions and identify potential areas for improvement in our thinking processes (Ellerton, 2022).

Integrating the Pedagogies through Meta-Cognitive

Meta-cognition can be integrated into various pedagogies to enhance the learning process for students. Here are some ways that meta-cognition can be integrated into different teaching pedagogies:

1. *Inquiry-Based Learning:* Inquiry-based learning involves encouraging the student to ask questions, investigate topics, and explore ideas. Meta-cognition can be integrated into this pedagogy by encouraging the students to reflect on own learning processes and identify areas where they need to improve their inquiry skills. Teachers can also help students develop meta-cognitive strategies such as setting goals, monitoring their own progress, and evaluating their own learning outcomes (Veenman and Kleyn, 2008).
2. *Problem-Based Learning:* It involves presenting students with real-world problems and challenging them to solve them using critical thinking and problem-solving skills. Meta-cognition can be integrated into this pedagogy by encouraging students to reflect on their own thinking processes and identify areas where they need to improve their problem-solving skills. Teachers can also help students develop meta-cognitive strategies such as analyzing the problem, breaking it down into smaller parts, and monitoring their own progress as they work towards a solution.
3. *Collaborative Learning:* It involves working in a group to solve problems or complete a project. Meta-cognition can be integrated into this pedagogy by encouraging students to reflect on their own learning processes within the group dynamic. Teachers can also help students develop meta-cognitive strategies such as monitoring their own progress within the group, identifying their own strengths and weaknesses, and evaluating their own contributions to the group (Jindal et al., 2016).
4. *Experiential Learning:* Experiential learning involves learning by doing, often through hands-on activities or real-world experiences. Meta-cognition can be integrated into this pedagogy by encouraging the students to reflect on own learning processes as they engage in the

experiential activity. Teachers can also help students develop meta-cognitive strategies such as setting goals, monitoring their own progress, and evaluating their own learning outcomes.

Integrating meta-cognition into different pedagogies can help students become more self-directed learners, develop critical thinking & problem-solving skills, and enhance their overall learning outcomes. By encouraging the students to reflect on own learning processes, teachers can help students develop the meta-cognitive strategies they need to succeed in school and beyond.

Integrating the Pedagogies through Critical Thinking Skills

Critical thinking skills can be integrated into various pedagogies to enhance the learning process for students. Here are some ways that critical thinking can be integrated into different teaching pedagogies:

1. *Inquiry-Based Learning:* It involves encouraging students to ask questions, investigate topics, and explore ideas. Critical thinking can be integrated into this pedagogy by encouraging students to analyze and evaluate information from multiple sources to draw their own conclusions. Teachers can also help students develop critical thinking skills such as identifying biases and assumptions, analyzing evidence, and evaluating arguments (Ennis, 2018).
2. *Problem-Based Learning:* It involves presenting students with real-world problems and challenging them to solve them using critical thinking and problem-solving skills. Critical thinking can be integrated into this pedagogy by encouraging students to analyze and evaluate the problem from multiple perspectives, brainstorm potential solutions, and evaluate the strengths and weaknesses of each solution.
3. *Collaborative Learning:* It involves working in groups to solve problems or complete projects. Critical thinking can be integrated into this pedagogy by encouraging students to analyze and evaluate the ideas and perspectives of their group members, and to work together to develop creative and effective solutions to problems.
4. *Experiential Learning:* Experiential learning involves learning by doing, often through hands-on activities or real-world experiences. Critical thinking can be integrated into this pedagogy by encouraging

students to analyze and evaluate the results of their actions and to reflect on how they can apply what they have learned to new situations.

Integrating critical thinking into different pedagogies can help students develop a more analytical and informed approach to learning, problem-solving, and decision-making. By teaching students to analyze and evaluate information, identify biases and assumptions, and develop creative solutions to problems, teachers can help students become more independent, self-directed learners, capable of succeeding in a wide range of academic and professional contexts (Ericson, 2022).

Inculcate Innovations and Connections through Meta-Cognitive

Meta-cognition can be a powerful tool for inculcating innovation and making connections between different ideas and concepts. Here are some ways that meta-cognitive strategies can help foster innovation and connections:

1. *Reflection:* Reflection is a critical component of meta-cognition and involves thinking about one's thinking processes. By encouraging students to reflect on their own learning experiences and identify patterns and connections between different concepts, teachers can help students make connections between seemingly disparate ideas.
2. *Goal Setting:* Goal setting is another meta-cognitive strategy that helps students inculcate innovation and make connections. By setting specific and challenging goals, students are encouraged to think creatively and innovatively to achieve those goals. Additionally, by setting goals that are interconnected and build upon each other, students are encouraged to make connections between different ideas and concepts.
3. *Self-Regulation:* Self-regulation is the ability to monitor one's thinking processes and adjust them as necessary. By teaching students to self-regulate their own thinking processes, teachers can help them identify when they are stuck in a rut or need to approach a problem from a new angle. This can encourage students to think creatively and come up with innovative solutions to problems.

4. *Metaphors and Analogies:* Metaphors and analogies can be powerful tools for making connections between different ideas and concepts. By encouraging students to use metaphors and analogies to explain complex ideas and make connections between different concepts, teachers can help them develop their own innovative approaches to learning and problem-solving (Šafranj, 2019).

By encouraging students to develop meta-cognitive strategies such as reflection, goal-setting, self-regulation, and the use of metaphors and analogies, teachers can help students inculcate innovation and make connections between seemingly disparate ideas. By promoting this kind of thinking, students can become more independent, self-directed learners, capable of succeeding in a wide range of academic and professional contexts (Sengupta and Blessinger, 2022).

Inculcate Innovations and Connections through Critical Thinking Skills

Critical thinking skills can be a powerful tool for inculcating innovation and making connections between different ideas and concepts. Here are some ways that critical thinking skills can help foster innovation and connections:

1. *Analysis:* Critical thinking involves analyzing information from multiple sources, identifying patterns and connections, and evaluating the strengths and weaknesses of different arguments. By teaching students to analyze information in this way, teachers can help them identify innovative solutions to problems and make connections between seemingly disparate ideas.
2. *Creativity:* Critical thinking also involves creativity - the ability to think outside the box and then come up with innovative solutions to various problems (Malik et al., 2021). By encouraging students to approach problems from multiple angles and consider unconventional solutions, teachers can help them develop their own innovative approaches to learning and problem-solving.
3. *Synthesis:* Critical thinking involves synthesizing information from multiple sources and drawing connections between different concepts. By teaching students to synthesize information in this way,

teachers can help them identify new and innovative ways to approach problems.

4. *Evaluation:* Critical thinking also involves evaluating the effectiveness of different solutions and approaches to problems. By encouraging students to evaluate the strengths and weaknesses of different arguments and solutions, teachers can help them identify innovative approaches to problem-solving (Kroeger et al., 2020).

By teaching students critical thinking skills such as analysis, creativity, synthesis, and evaluation, teachers can help them inculcate innovation and make connections between seemingly disparate ideas. By promoting this kind of thinking, students can become more independent, self-directed learners, capable of succeeding in a wide range of academic and professional contexts.

Role of Meta-Cognitive and Critical Thinking Skills in Problem-Solving

Meta-cognition refers to the ability to think about one's thinking process. This involves understanding how we acquire, process, and use information to solve problems and make decisions. By being aware of our thought processes, we can monitor and adjust them to improve our problem-solving and decision-making skills.

Critical thinking is analyzing information, evaluating arguments, and drawing conclusions based on evidence. It involves questioning assumptions, identifying biases, and considering alternative viewpoints. We can make informed decisions and solve complex problems by using critical thinking skills.

Meta-cognitive and critical thinking skills are crucial in problem-solving and decision-making because they enable us to.

1. *Identify the problem or decision to be made:* Meta-cognitive skills help us recognize when we are faced with a problem or decision and determine what needs to be done to solve it.
2. *Analyze the situation:* Critical thinking skills help us to break down complex problems into small parts, identify relevant information, and evaluate its accuracy and reliability.

3. *Develop a strategy:* Meta-cognitive skills help us plan and organize our thoughts and actions, while critical thinking skills help us identify the best course of action.
4. *Evaluate the solution:* Meta-cognitive skills help us reflect on the effectiveness of the solution, while critical thinking skills help us assess its strengths and weaknesses and identify ways to improve it.

Meta-cognitive and critical thinking skills are essential in problem-solving & decision-making because they enable us to approach complex problems and decisions systematically and logically, leading to more effective and informed outcomes.

Role of Meta-Cognitive in Skill and Reskill Development

Meta-cognition can be crucial in skill and reskill development by helping individuals become more aware of their own learning processes and strategies. When we engage in meta-cognitive practices, we can identify what works and what doesn't work for us when learning new skills, and we can adjust our approach accordingly.

Here are some ways in which meta-cognition can help in skill and reskill development-

1. *Planning and goal-setting:* Meta-cognition can help individuals set specific and achievable goals for their skill development. By reflecting on their current skills and identifying areas for improvement, individuals can set goals that are realistic and relevant to their needs.
2. *Self-monitoring:* Meta-cognition can help individuals monitor their progress toward their skill development goals. By regularly reflecting on their learning experiences, individuals can identify what is working and what is not working and make adjustments as needed.
3. *Feedback and reflection:* Meta-cognition can help individuals seek and use feedback from others to improve their skills. By reflecting on feedback and considering how to apply it to their learning process, individuals can make more informed decisions about how to develop their skills.
4. *Motivation and persistence:* Meta-cognition can help individuals stay motivated and persistent in their skill development efforts. By

reflecting on their progress and celebrating successes, individuals can maintain a positive attitude and stay committed to their goals.

Meta-cognition can be a powerful tool in skill and reskill development by helping individuals become more aware of their own learning processes and strategies, and by enabling them to make more informed decisions about how to develop their skills.

Role of Critical Thinking in Skill and Reskill Development

Critical thinking can be a valuable asset in skill and reskill development by helping individuals assess and analyze information, evaluate different options, and make informed decisions about how to develop their skills (Kallet, 2014). Here are some ways in which critical thinking can help in skill and reskill development:

1. *Identifying relevant information:* Critical thinking can help individuals identify and evaluate the information they need to develop their skills effectively. By assessing the credibility and relevance of different sources of information, individuals can make informed decisions about what to learn and how to learn it.
2. *Analyzing options:* Critical thinking can help individuals assess and compare different options for developing their skills. By weighing the pros and cons of different approaches, individuals can make informed decisions about which methods will be most effective for them (McMillan and Weyers, 2013).
3. *Problem-solving:* Critical thinking can help individuals identify and solve problems that arise in the process of developing their skills. By breaking down complex problems into smaller parts and evaluating different solutions, individuals can make informed decisions about how to overcome obstacles and continue making progress.
4. *Evaluating outcomes:* Critical thinking can help individuals evaluate the outcomes of their skill development efforts. By assessing the effectiveness of different methods and reflecting on what worked and what didn't, individuals can make informed decisions about how to refine their approach going forward.

Critical thinking can be an invaluable tool in skill and reskill development, helping individuals make informed decisions about what to learn, how to learn it, and how to evaluate and refine their approach over time (Farcis et al., 2021).

Conclusion

STEM is a multi-disciplinary approach that integrates knowledge and skills from different fields to solve complex problems and drive innovation. It encourages students to think creatively, critically, and analytically to tackle real-world challenges and prepare them for success in a rapidly changing world. Meta-cognitive and critical thinking skills are two important cognitive abilities that play a crucial role in our learning, problem-solving, and decision-making processes. By developing these skills, we can become more self-directed learners, make informed decisions, and effectively communicate our ideas and perspectives. Meta-cognitive and critical thinking skills are mutually reinforcing, as they both involve the ability to analyze and evaluate information, reflect on one's own thinking processes, and make informed judgments and decisions. By developing both of these skills, we can enhance our ability to learn, problem-solve, and make informed decisions in all areas of life.

References

Ahmed, A. A. A., Keezhatta, M. S., Khair Amal, B., Sharma, S., Shanan, A. J., Ali, M. H., & Farooq Haidari, M. M. (2022). The comparative effect of online instruction, flipped instruction, and traditional instruction on developing Iranian EFL learners' vocabulary knowledge. *Education Research International*, *2022*, 9.

Anderson, J., & Li, Y. (2020). Investigating the potential of integrated STEM education from an international perspective. *Integrated Approaches to STEM Education: An International Perspective,* 1-12.

Bradley, S., & Price, N. (2021). *Critical thinking: Proven strategies to improve decision making skills, increase intuition and think smarter.* Createspace Independent Pub.

Chen, J. (2022). *Cognitive Mapping for Problem-based and Inquiry Learning: Theory, Research, and Assessment.* London, UK: Routledge.

Ellerton, P. (2022). On critical thinking and content knowledge: A critique of the assumptions of cognitive load theory. *Thinking Skills and Creativity*, 43, 100975.

Ennis, R. H. (2018). Critical thinking across the curriculum: A vision. *Topoi,* 37, 165-184.

Ericson, J. D. (2022). Mapping the relationship between critical thinking and design thinking. *Journal of the Knowledge Economy*, 13(1), 406-429.

Farcis, F., Wijayanti, E., Budi, G. S., & Dinata, P. A. C. (2021). Critical Thinking Skills And Student Enthusiasm In Virtual-Based Learning For Thermodynamics Course. *Jurnal Ilmiah Kanderang Tingang,* 12(02), 205-216.

Johnson, C. C., Mohr-Schroeder, M. J., Moore, T. J., & English, L. D. (2020*). Handbook of research on STEM education,* 526. London, UK: Routledge.

Jindal, P., & Tej, F. (2016). Being an educator in private vs. public school system: A comparative study on job contentment. *IRA International Journal of Education and Multidisciplinary Studies*, *3*(3), 309-319.

Kumari, S., Jindal, P., & Chand, P. K. (2022). Social Media Networking: As Facilitator of Acculturation. *ECS Transactions*, *107*(1), 9381.

Kroeger, S. D., Howley, A., Hansen, B., & Faiella, C. M. (2020). Restructuring Teacher Education Programs for Inclusive Instructional Practice. *Inclusive Education: A Systematic Perspective*, 191.

Kallet, M. (2014). *Think smarter: critical thinking to improve problem-solving and decision-making skills.* John Wiley & Sons.

MacDonald, A., Danaia, L., & Murphy, S. (2020). *STEM Education Across the Learning Continuum.* Singapore: Springer.

Malik, R., Shrama, A., Trivedi, S., & Mishra, R. (2021). Adoption of Chatbots for learning among university students: role of perceived convenience and enhanced performance. *International Journal of Emerging Technologies in Learning (iJET)*, 16(18), 200-212.

McMillan, K., & Weyers, J. (2013). *How to improve your critical thinking & reflective skills.* Pearson Higher Education.

Penprase, B. E. (2020). *STEM Education for the 21st Century.* Springer Nature.

Šafranj, J. (2019). The effect of meta-cognitive strategies on self-efficacy and locus of control of gifted in foreign language learning. *Research in Pedagogy*, 9(1), 40-51.

Sengupta, E., & Blessinger, P. (Eds.). (2022). *ICT and Innovation in Teaching Learning Methods in Higher Education*. Emerald Group Publishing.

Veenman, M., & Kleyn-Kennedy, C. (2008). Meta-cognition: A recent review of research, theory, and perspectives. *Meta-cognition: A recent review of research, theory, and perspectives,* 207-220.

Wilson, D., & Conyers, M. (2016). *Teaching students to drive their brains: Metacognitive strategies, activities, and lesson ideas.* ASCD.

Zhou, J., & Wang, X. (2020). The influence of chinese college students' meta-cognitive strategy and motivational beliefs on their study habits. *The Asia-Pacific Education Researcher*, 1-11.

Chapter 6

STEM: A Collaborative, Cooperative and Skill-Based Learning

Kajal Rani[1,*]
and Priya Jindal[2]**, PhD**
[1]Chandigarh Vidya Jyoti Eduversity, Chandigarh, India
[2]Chitkara Business School, Chitkara University, Punjab, India

Abstract

Cooperative learning is working together in a small group in a structured activity. The individual is accountable for their work; students are arranged in groups. In collaborative learning, two or more people learn something together. In these face-to-face conversations occur, people collaborate. Students work together to search for understanding. In skill-based learning, education emphasizes developing practical knowledge rather than a book, developing practical skills for students that they apply in real life. Skill-based learning helps students increase their learning capability and allows them to handle the real world. In skill-based learning, students are pushed to think outside the box and innovate some newly developed new technology. This student becomes a quicker learner; some schools organize skill development programs to prepare students for their future, help them become academically strong, and build the capability to deal with situations.

Keywords: cooperative learning, collaborative learning, skill-based learning, practical knowledge

---

* Corresponding Author's Email: kajal76588@gmail.com.

In: STEM
Editors: Priya Jindal, Amandeep Singh and Rashmi Aggarwal
ISBN: 979-8-89113-305-1

Introduction

STEM stands for science, technology, engineering, and math and is frequently used when talking about school curriculum, academic discipline, and educational policy. Students collaborate in groups as part of the teaching and learning process known as collaborative learning to explore an important question or create a noteworthy project. One example of collaborative learning is when students from different schools work together online to finish a group project.

Cooperative and Collaborative Learning Have Commonalities

Few examples of how cooperative learning and collaborative learning strategies for students are similar:

1. No passive lecture, lecture based on the active participation of students.
2. Students have active involvement in education.
3. Teachers facilitate learning by doing so.
4. Make a task or project mandatory for students to develop teamwork skills and promote social interaction.

Table 1 shows the difference between cooperative and collaborative learning. Collaborative learning is a group-structured mode of learning where each student achieves self-development progressing and flourishing simultaneously as a team, while cooperative learning is a group-structured mode of learning where the members of the student group have the responsibility of enhancing the academic performance of the others.

Table 1. Difference between cooperative and collaborative learning

Cooperative Learning	Collaborative Learning
Teacher-centered	Student-centered
Limited responsibility	Participants work together to solve a problem
Different goals	Common goals
Closed assignments	Open assignments
Static	Dynamic

Source: Babić et al., (2021).

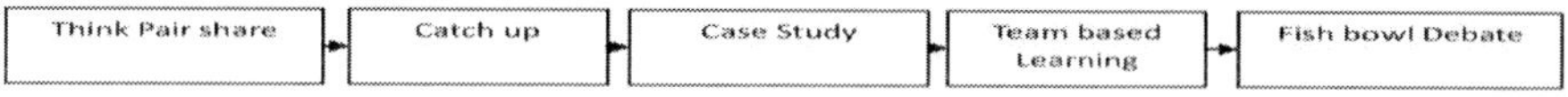

Source: www.keydifferences.com.

Figure 1. Collaborative learning strategies.

Students engage in cooperative learning when they collaborate in small groups to accomplish a common objective. Figure 1 shows the different strategies of collaborative learning that may be used by teachers in any grade.

Skill-Based Learning

It is an approach to education that helps develop practical knowledge rather than book learning, and students apply this knowledge in real-world situations rather than just acquiring knowledge. Skill-based learning helps students tap into their capabilities and prepares them for the future. The following are the advantages of Skill-based education:

1. **Help in acquiring practical knowledge:** Skill-based education helps to gain valuable understanding rather than book knowledge. They help the student to think outside of the box.
2. **Developing essential skills:** STEM-based learning skills are crucial for students so that the student may apply them in real life.
3. **Student gains more profound knowledge:** Skill-based education provides practical knowledge, so students get a deep understanding of how they happen, why, what kind of reason behind this, etc. This type of knowledge students can gain through skill-based education.
4. **Encourage students to actively engage:** Skill-based education provides practical knowledge, so it increases students' interest levels. Student involvement is high.
5. **Enhance social and emotional well-being:** They help the student manage their emotions, make responsible decisions, build and maintain relationships, and respect and understand another person's emotions.
6. **Enhance student thinking capability:** Skill-based learning helps students to solve problems, make different and accurate decisions, make a proper plan, and organize and create information.

Example of Skill-based Learning

Examples of skill-based learning are projects in which students collaborate, take responsibility, and develop teamwork, problem-solving, critical thinking, etc.

Importance of Skill-Based Curriculum in Higher Education

1. **Help in experience:** A skill-based curriculum helps in higher education to explore possible career options. Increase self-confidence and self-understanding and help young people for work.
2. **Promote thinking ability:** They help the student solve the problem, make a different decision, and enhance thinking ability.
3. **Help in growth:** Skill-based education in higher education helps students for something progress and become more successful.
4. **Increase knowledge:** Skill-based education provides practical knowledge to help students enhance their knowledge and learn something new and innovative.
5. **Promote advanced training:** They help the students to handle more duties and responsibilities for promotion to their higher job in the organization and achieve their own goals.
6. **Help in competence development:** Developing competence is identifying the skills, talents, and knowledge the student requires to perform the job effectively and training to practice to improve them.
7. **Job Readiness:** Graduates with the requisite abilities for the position are frequently sought after by employers. A skill-based curriculum aids students in developing the abilities necessary to succeed in the profession by emphasizing practical skills including critical thinking, communication, teamwork, problem-solving, and decision-making.
8. **Real-world application:** A skill-based curriculum places a strong emphasis on knowledge's practical application. Learning by doing aids students in applying their skills and knowledge to issues in the real world. This not only improves their education but also gets them ready for the difficulties they will encounter at work.
9. **Relevant and up-to-date:** Compared to standard academic programs, skill-based curricula are more relevant and up-to-date. It is made to adapt to the shifting needs of the labor market and give students the abilities they require to thrive in the twenty-first century.

10. **Learning that is personalized:** A skill-based curriculum enables students to concentrate on their areas of interest and acquire abilities that are particular to their professional aspirations. This allows students to take charge of their education and follow their passions.
11. **Continuous learning:** Lifelong learning is promoted via skill-based curricula. Even after they graduate from college, students are urged to keep growing intellectually and acquire new talents. This is significant at a time when innovation in terms of technologies and methods is accelerating at a rapid rate.

Collaborative Learning in STEM

The proposed chapter tries to ascertain how collaborative learning affected students' attitudes about the course material and behavior connected to it in Duke undergraduate STEM courses. This Bass Connections team looked studied whether collaborative learning increases underrepresented student populations' sense of self-efficacy and motivation to stick with STEM disciplines. To assess the effects of the STEM course format on course-related attitudes and learning outcomes, the team developed a survey.

According to their findings, students in classrooms with more structure demonstrated:

1. An increase in engagement
2. Increased knowledge of the facts
3. Greater assurance in your content knowledge
4. Increased capacity for knowledge application, analysis, and synthesis
5. A stronger sense of kinship
6. More of a sense of belonging.

Review of Literature

Although the Currency, Relevance, Authority, Accuracy, and Purpose (CRAAP) test typically lacks a C &C element, we use cooperative learning in this approach. This task incorporates a group discussion and student interaction. In groups of two to three students, they first select an article on a related subject, which they then read. Students argue over whether the referenced source met the CRAAP requirements or not, as well as whether it is relevant and trustworthy. The articles are then discussed in-depth in front of the class, whether they pass the CRAAP test, among other things. According to the dual-coding hypothesis suggested by Brown and Palincsar (2018), multimedia is thought to improve learning because pupils receive auditory and visual information concurrently, improving retention. Collaborative learning aided multimedia and the web due to the continuing infusion of digital resources into higher education (Davidson and Major, 2014).

Chi (2009) tried cooperative learning exercises using video as a multimedia tool for learning psychological topics in our developmental psychology classes. Videos can make abstract concepts clear and memorable when appropriately used. Videos can also illustrate psychological ideas that can be used as a starting point for debate and practical application of course material (Thibaut et al., 2018). The "activity" of A video could encourage passive learning as well unless it is specifically designed to engage students. The use of literature and multimedia should be deliberate and active (Herrmann, 2013). We set out to develop a collection of lectures focused on group discussion of movies depicting conventional developmental scenarios and research using these warnings in mind. To teach fundamental developmental principles (such as object permanence, joint attention, attachment, and egocentrism), we collaborated on the production of PowerPoint lesson modules with connected videos with three other graduate student developmental educators who were GSTA members. The collaborative course prepares for human development in which several of the Graduate Student Teaching Association (GSTA) instructors from different campuses within the university system share instructional materials, making using this medium much simpler (Salomone and Kling, 2017). Each module included two movies carefully chosen from YouTube and showed a developmental experiment or circumstance.

STEM Skills

Problem-solving, critical thinking, creativity, and reasoning are among the STEM skills. They may be learned and improved with practice, much like other abilities. While some STEM talents fall under the category of soft skills, some might be categorized as hard skills. The goal of STEM education is to incorporate these four academic fields into the classroom and to motivate students to use what they learn in practical settings. Students gain a deeper grasp of how the world functions and how to use scientific ideas to solve problems thanks to this multidisciplinary approach to learning.

Students often collaborate in teams to design, build, and test solutions to issues as part of their project-based, hands-on STEM education (Jindal and Tej, 2016). This strategy promotes problem-solving, analytical thinking, and creativity as well as teamwork and communication.

These five talents are the STEM fields emphasizing the development of students:

1. Problem-Solving

Many of the difficulties you'll encounter in STEM professions are technical issues that need to be solved using knowledge of scientific and mathematical principles. It necessitates a detailed examination of the numerous contributing elements and the application of scientific knowledge. The scientific method must be used while solving problems.

2. Critically Analysis

Without the use of critical thinking, great scientific discoveries are not conceivable. This ability involves determining if data or information is trustworthy. Even if a well-known expert presents an idea or concept, it must still be evaluated on its own merits without being swayed by the reputation or authority of the proponent.

3. Creativity Curiosity

Innovation and the capacity to perceive new approaches to problems are two characteristics of creativity. It also refers to more useful things, not just beauty. You may enhance systems and make them more dependable and efficient by using creativity. It results in discoveries and alternate routes. Children naturally possess a scientific curiosity, which is also a key quality of a scientific mindset and aptitude. Simply by learning how to ask the correct questions, you may enhance your capacity for curiosity. They're more likely to get the proper answers and ask the right questions. To ask the correct questions, we must first acquire all the necessary background information. The core of the study is this.

4. Mathematical and Logical Ability

Science operates under the presumption that the universe and its events are understandable and measurable to some extent. This implies that natural rules can be reduced to mathematical terms. Consider the Newtonian law of gravity as an example. Natural laws can only be represented by mathematical equations that can be empirically supported by quantifiable facts. The ability to establish logical connections and predictions regarding natural occurrences based on quantifiable facts is referred to as having logical-mathematical skills.

5. Abilities in Engineering Design

Additionally, designing and building things are part of STEM education. require engineering design abilities to help people comprehend the mathematical and scientific basis of designs, whether they are computer programs or mechanical or structural designs (Kumari et al., 2022). Designing mechanical and digital systems that one can manipulate or experiment with is made possible by STEM education.

How to Develop and Integrate STEM Education in Institutions and Schools

Many schools, universities, and nations with basic educational systems have specialized curricula for STEM education, but it is also possible to incorporate STEM education within the regular general curriculum. Any teacher, regardless of grade level, can include STEM education in the subjects they are teaching. A history teacher might spend time talking about the history of science, but an English language instructor might include science fiction among their reading resources.

A school should set aside funds and other resources to assist STEM education if it wants to advance it. For instance, the library should have the most recent editions of books, periodicals, and other materials about science and engineering. Additionally, there must be labs and areas for experiential learning.

STEM Education Goals and Objectives

Goal 1 - Making Use of Data and Insights

STEM education strives to equip students with knowledge and understanding of the scientific, technical, engineering, and mathematical fields. The idea enables students to investigate a variety of topics and projects.

Goal 2 - Application of Recommended Practises

In all of the STEM areas, students must be able to apply best practices. The long-term objective is to improve aptitude.

Goal 3 - Acquiring Knowledge and Being Creative

The pupils should learn about the four STEM fields and develop their capacity for innovation. Engineers might be asked to create an "L-shaped" wooden instrument in the lab, for instance, by their instructors. Here, the pupils are welcome to ask the lecturers for recommendations. Additionally, they might go see a carpenter to learn about the steps involved in making the tool, such as:

1. Discovering numerous topics
2. Choose the appropriate material
3. Performing project work
4. Angle cutting and filing
5. Completing the task.

Elements of STEM Education

Research-based Education

The students in this style of learning gather data, analyze it, develop scientific hypotheses, and pinpoint issues to carry out research tasks. They also create a prototype to address the issue before coming up with the best answer. Students can do this by using technology and mathematics to solve real-world issues.

Knowledgeable Education

It is a method of learning that is integrated into the lectures in a way that motivates the students to critically analyze their assumptions. It prevents people from studying to memorize and recollect knowledge for the tests.

Technology, Life Skills, and Career

Making the kids proficient in the subjects they are learning is part of this. It places a strong emphasis on project-based learning, internships, and practical experiences for students. The student's critical thinking, teamwork, communication, and creativity are all goals of the program.

Individualization of Instruction

A change from teacher-led classrooms to more student-centered learning is required for personalized learning. In this, the student has the option to select their subjects, learn at their own pace, and take tests as needed.

Connection to the External and Larger Community

Making the pupils aspirational and global-minded is necessary for this. Local environments and community organizations are made available to the students as learning tools. The children are aware of problems in their community and work to resolve them.

Following are Some Guidelines for Establishing STEM Education

1. The STEM program's aims and objectives should be determined. This includes establishing the outcomes that the program seeks to attain as well as the abilities and information that students should gain.
2. Create the curriculum under the aims and objectives of the program. This entails developing lesson plans, picking suitable teaching resources, and choosing evaluation techniques.
3. Give instructors access to professional development. To help teachers adjust to the new teaching strategy, which differs from conventional approaches for STEM education, it is crucial to offer training and support.
4. Form alliances with companies and groups working in STEM-related industries. These collaborations may offer students the chance to

work on real-world projects and acquire practical experience for STEM-related employment.

5. Include resources and technology in the curriculum. Technology and other resources are heavily incorporated into STEM education to give students a dynamic and engaging learning environment.
6. Encourage transdisciplinary study. Instead of focusing only on one subject, STEM education should encourage interdisciplinary study so that students can understand the links between them and use what they have learned in a variety of contexts.
7. Review the plan and make any necessary adjustments. To guarantee that the program continues to fulfill the needs of students and achieve its aims and objectives, regular evaluation and assessment can assist in identifying areas that need improvement and enable improvements to be made to the program.

Benefits of Using Cooperative Learning, Collaborative Learning, and Hybrid Cooperative-Collaborative Techniques

- Importance of collaborative and cooperative teaching strategies
- Enhancing students' classroom experiences and interactions with other students
- Collaborating with other instructors and advisory faculty to develop course preparation

STEM (Science, Technology, Engineering, and Mathematics) learning offers numerous benefits for students. By integrating these disciplines, students develop a holistic understanding of real-world problem-solving and critical thinking skills. STEM education fosters creativity, innovation, and collaboration, preparing students for the demands of the modern workforce. Furthermore, STEM learning promotes curiosity, inquiry-based learning, and a growth mindset, encouraging students to explore and experiment with ideas. As society continues to advance technologically, STEM education equips students with the knowledge and skills needed to thrive in a rapidly changing world (Tuli et al., 2017). Overall, STEM education fosters lifelong learners and equips students to participate fully in a society that places an increasing emphasis on STEM fields. Collaborative learning is a potent educational strategy that improves students' learning opportunities and social skills.

Students learn effective cooperation, leadership, and communication skills through cooperating in groups. Active participation is encouraged by collaborative learning because it allows students to exchange different viewpoints, learn from and educate one another, and work together to solve challenges (Faridi et al., 2021). Collaboration and teamwork are widely appreciated in both the personal and professional worlds, so collaborative learning not only enhances academic results but also equips students with the skills they need to succeed in both.

Limitations of STEM Education

A Narrow Focus

Focusing too narrowly on science, technology, engineering, and math might result in a STEM education that neglects other crucial subjects like language arts, social studies, and the humanities. This may lead to a narrow view of crucial issues and a lack of a well-rounded education. STEM education places more of an emphasis on learning technical skills, such as coding or laboratory procedures, than on building soft skills. However, it might not place as much focus on the improvement of soft skills like leadership, teamwork, and communication. As a result, graduates may have a solid technical background but lack the interpersonal skills required for professional success.

Lack of Diversity

White men have historically held the majority of positions in STEM professions, and STEM education may not be doing enough to address this issue. This may lead to a lack of opportunities and representation for underrepresented groups, which may result in the loss of talent and viewpoints.

High Resource Needs

STEM education frequently calls for specialized tools, materials, and resources, which can be costly and challenging to procure, especially for schools with tight budgets. This may lead to unequal access to opportunities and high-quality STEM education.

Prioritise Memorization over Originality

Memorization is prioritized above creativity in STEM education, which can occasionally favor rote learning and memorization over innovation and problem-solving. Students who are good at following instructions but not creative or critical thinkers may be the outcome of this.

Training for Instructors

In schools with few resources, many teachers do not have the knowledge and skills essential to teach STEM subjects well. As a result, kids may receive poor instruction and miss out on possibilities.

Lack of Application to Real-World Issues

Some STEM education initiatives may fail to adequately apply the concepts and abilities students have gained in the classroom to real-world issues and applications. This may lead to students who find it difficult to understand the value of their education and who may not be well-prepared for the workforce.

Limited Career Alternatives

If students are not interested in pursuing professions in typical STEM sectors, STEM education may not offer them a wide enough range of employment options. As a result, students may have fewer employment options and less freedom to alter their career routes in the future.

Bias in Favour of Quantitative Methods

STEM education frequently places a focus on quantitative methods of problem-solving, including data analysis and mathematical modeling. However, some issues might call for more qualitative methods, such as ethnographic study or observation. In some situations, this bias may impede pupils' ability to solve problems.

Lack of Attention to Societal Implications

STEM education may not devote enough time to discussing the moral and societal ramifications of scientific and technical advancements. Graduates from such programs might be technically proficient but unable to assess critically the possible effects of their work on society and the environment.

Overall, even though STEM education has numerous advantages, it is crucial to take these drawbacks into account and seek to overcome them to offer all children a high-quality, inclusive, and successful STEM education.

Conclusion

The benefits of using the teaching of psychology, cooperative learning, collaborative learning, and hybrid cooperative-collaborative techniques have been highlighted in this chapter. They can improve the motivation and learning of students. The course assignments can be changed in a variety of ways to incorporate group-based learning we highly recommend it Try these collaborative and cooperative teaching strategies in the classroom. Kids find cooperative and collaborative work very stimulating, and it also helps kids to develop essential skills like communication and critical thinking. Students' classroom experiences and interactions with other students are enhanced when they enjoy their learning. Improves the possibility of using their newly acquired knowledge and growing skills for upcoming group projects. Last, those who benefited from teamwork in our classes were not just the pupils. We collaborated with other instructors and advisory faculty to develop our course preparation, sharing resources and ideas.

References

Brown, A. L., & Palincsar, A. S. (2018). Guided, cooperative learning and individual knowledge acquisition. In *Knowing, learning, and instruction*, 393-451. Routledge.

Babić, T., Kolar, L., & Miličević, M. (2021). Individual, Cooperative and Collaborative Learning and Students' Perceptions of Their Impact on Their Own Study Performance.

In *2021 44th International Convention on Information, Communication and Electronic Technology (MIPRO)*, 864-869. IEEE.

Chi, M. T. (2009). H., 2009. *Active-Constructive-Interactive: A Conceptual Framework for Differentiating Learning Activities, Topics in Cognitive Science*, *1*, 73-105.

Davidson, N., & Major, C. H. (2014). Boundary crossings: Cooperative learning, collaborative learning, and problem-based learning. *Journal on excellence in college teaching*, *25*.

Faridi, H., Tuli, N., Mantri, A., Singh, G., & Gargrish, S. (2021). A framework utilizing augmented reality to improve critical thinking ability and learning gain of the students in Physics. *Computer Applications in Engineering Education*, *29*(1), 258-273.

Herrmann, K. J. (2013). The impact of cooperative learning on student engagement: Results from an intervention. *Active learning in higher education*, *14*(3), 175-187.

Jindal, P., & Tej, F. (2016). Being an educator in private vs. public school system: A comparative study on job contentment. *IRA International Journal of Education and Multidisciplinary Studies*, *3*(3), 309-319.

Kumari, S., Jindal, P., & Chand, P. K. (2022). Social Media Networking: As Facilitator of Acculturation. *ECS Transactions*, *107*(1), 9381.

Salomone, M., & Kling, T. (2017). Required peer-cooperative learning improves retention of STEM majors. *International Journal of STEM Education*, *4*(1), 1-12.

Shimazoe, J., & Aldrich, H. (2010). Group work can be gratifying: Understanding & overcoming resistance to cooperative learning. *College teaching*, *58*(2), 52-57.

Surbhi, S. (2021). Difference Between Collaborative Learning and Cooperative Learning. Retrieved from https://keydifferences.com/difference-between-collaborative-learning-and-cooperative-learning.html.

Thibaut, L., Ceuppens, S., De Loof, H., De Meester, J., Goovaerts, L., Struyf, A., Boeve-de Pauw, J., Dehaene, W., Deprez, J., De Cock, M., Hellinckx, L., Knipprath, H., Langie, G., Struyven, K., Van de Velde, D., Van Petegem, P. and Depaepe, F. (2. (2018). Integrated STEM education: A systematic review of instructional practices in secondary education. *European Journal of STEM Education*, *3*(1), 2.

Tuli, N., Gargrish, S., & Mantri, A. (2017, December). Low-cost learning environment for grass-root education. In *2017 IEEE Region 10 Humanitarian Technology Conference (R10-HTC)*, 615-619. IEEE.

Chapter 7

The Application of Meta-Cognitive, Critical Thinking, Collaborative, Cooperative and Skill Based Learning Strategies in STEM

Natasha Pallavi*
Faculty Associate, ICFAI Tech School-The ICFAI University, Dehradun, India

Abstract

The main emphasis of the chapter is on various learning strategies in STEM that promote meta-cognition, critical thinking, collaboration, cooperation, and skill development. These diverse approaches are vital aspects of comprehensive education and can be applied in different educational settings. Meta-cognitive learning involves comprehending one's own learning and thinking processes, including awareness of effective strategies for enhancing learning. Critical thinking entails logically and objectively analyzing and evaluating information, enabling learners to make well-informed decisions based on evidence and reasoning. Collaborative learning in STEM involves working collectively towards a shared objective, fostering communication, cooperation, and the exchange of knowledge and skills. Cooperative learning, similar to collaborative learning, places importance on individual accountability and the responsibility to contribute to the group's overall success. Skill-based learning concentrates on honing specific competencies like problem-solving, communication, and leadership skills. These proficiencies are crucial for success in the workforce and everyday life. By incorporating these diverse learning approaches into STEM education, students' comprehension and retention

* Corresponding Author's Email: natasha.pallavi@iudehradun.edu.in.

In: STEM
Editors: Priya Jindal, Amandeep Singh and Rashmi Aggarwal
ISBN: 979-8-89113-305-1

of information can be enhanced. Furthermore, their problem-solving and decision-making abilities can be improved, preparing them for success in the 21st-century job market.

Keywords: meta-cognitive, critical thinking, collaborative, cooperative, skill, STEM

Introduction

Meta-cognitive, critical thinking, collaborative, cooperative, and skill-based learning are all important approaches to learning that can help students develop the necessary knowledge, skills, and attitudes to succeed in their academic and professional lives. In this chapter, the author will explore each of these learning approaches and discuss their key features, benefits, and challenges.

In this chapter, the author will explore each of these learning approaches and discuss their key features, benefits, and challenges. Meta-cognitive learning involves teaching students how to think about their thinking, or in other words, how to become more aware of their thought processes and strategies for learning. This approach emphasizes the importance of self-reflection, self-monitoring, and self-regulation in the learning process. By developing these skills, students can better identify their strengths and weaknesses, set goals, and develop effective study strategies. Meta-cognitive learning has been shown to improve students' academic performance, increase their motivation, and enhance their overall learning experience.

Metacognition is the ability to think about one's thinking. It involves understanding how we learn, what strategies work best for us, and how we can monitor our own thinking and learning processes. Metacognition enables learners to take control of their learning by setting goals, monitoring progress, and adjusting strategies as needed. Metacognitive skills include self-awareness, self-reflection, planning, monitoring, and evaluating. For example, a metacognitive student might set a goal to learn a new concept, plan how they will approach the learning process, monitor their understanding as they learn, reflect on what strategies were effective, and evaluate their overall learning. Incorporating metacognitive strategies into learning can enhance students' understanding and retention of information, as well as their ability to transfer what they have learned to new situations. It can also improve students' confidence and motivation by giving them a sense of control over their learning. Meta-cognition plays a crucial role in learning. By understanding their learning processes and strategies, learners can identify their strengths and weaknesses and adjust their approach to learning accordingly. For example, a student who struggles with memorization might realize that they learn best by visualizing concepts, and therefore use more visual aids to support their learning.

In addition, meta-cognitive strategies such as planning, monitoring, and evaluating can help learners become more effective and efficient in their learning. For example, a student might plan out their study schedule and set goals for each study session, monitor their progress throughout the session to ensure they are staying on track, and evaluate their learning at the end to identify areas that need further review.

Meta-cognition can also help learners transfer their learning to new situations. By understanding the underlying concepts and principles, learners can apply what they have learned to new problems or situations. This transfer of learning is a key aspect of education, as it enables learners to apply their knowledge and skills in real-world contexts. Overall, incorporating meta-cognitive strategies into learning can enhance learners' understanding and retention of information, improve their problem-solving and decision-making abilities, and prepare them for success in the 21st-century workforce.

Meta-Cognition and its Benefits

Meta-cognition refers to the ability to think about one's thinking processes. It involves being aware of and understanding one's thought processes, such as how one learns, remembers, solves problems, and makes decisions. Meta-cognition also involves being able to monitor and regulate one's cognitive processes, such as attention, memory, and motivation.

The benefits of meta-cognition are numerous, including:

- *Improved Learning*: By understanding their learning processes, individuals can become more efficient and effective learners. They can identify the best strategies to use when learning new information, monitor their understanding, and adjust their learning strategies accordingly.
- *Better Decision-Making*: Meta-cognition can help individuals make better decisions by enabling them to reflect on their decision-making

processes. They can identify any biases or assumptions that may be influencing their decisions and make more informed choices.

- *Enhanced Problem-Solving*: Meta-cognition can also aid in problem-solving by helping individuals recognize when they are stuck and identify the best strategies to use to solve a particular problem.
- *Increased Self-Awareness:* By reflecting on their thought processes, individuals can become more self-aware and better understand their strengths and weaknesses. This can help them to identify areas for improvement and work towards self-improvement.
- *Improved Communication*: Meta-cognition can help individuals to communicate more effectively by enabling them to better understand their thought processes and by providing them with the tools to articulate their thoughts and ideas more clearly.

Overall, meta-cognition is a valuable skill that can help individuals to improve their cognitive processes, make better decisions, and become more effective learners and problem-solvers.

Literature Review in STEM Education

The application of meta-cognition in learning enhances the skills of learners and also improves the content knowledge of teachers by encouraging them to think about their thinking (Chauhan et al., 2021). Basic cognitive knowledge is necessary to achieve a specific goal (Ekici, 2021). Ismail et al., (2018) suggest that the strategy category includes a significant amount of knowledge that can be acquired depending on the effectiveness of skills and strategies in problem-solving during learning. To better understand the application of meta-cognition in learning skills and strategies that contribute to the achievement of goals or problem-solving, it is important to explain the concept of learning skills and strategies used in this study. "Learning skills" refers to proficient activities, methods, or expertise that assist learners in problem-solving, while "learning strategies" refers to behaviors or actions that learners employ to make mathematics learning in problem-solving more successful, self-directed, and enjoyable (Perry et al., 2019).

In Singapore, meta-cognition has been recognized as one of the key components essential for the development of mathematical problem-solving in the mathematics curriculum (Wiggins, 2012). According to Roschelle and

Teasley (1995), meta-cognitive strategies encompass a set of competencies for learning and thinking, which include the skills required for active learning, critical thinking, reflective judgment, problem-solving, and decision-making. In other words, learners who possess well-developed meta-cognitive skills, achieved through thinking about their thinking, are better equipped to solve problems, make decisions, and think critically, all of which enhance their learning skills. Therefore, meta-cognition is a crucial aspect of student learning, and the development of meta-cognitive skills and strategies enables students to better organize their thought processes and refine their problem-solving abilities (Bakar and Ismail, 2020). Facione (2011) asserts that basic cognitive knowledge is necessary to accomplish a specific goal, as it guides in a learning situation. Additionally, the appropriate use of meta-cognitive skills and strategies helps individuals acquire knowledge and intellectual abilities that guide and direct their intellectual processes in learning (Johnson and Johnson, 1987). The metacognition process assists learners in effectively communicating the steps taken to solve a problem. Learners should be able to explain why those steps are valid in executing the action, particularly in conceptual reasoning with mathematics or the problem-solving process.

When learners reflect on their achievements by thinking about their thinking, it can be seen as a component of their higher-order reasoning and reflective questioning, both of which contribute to the improvement of their learning abilities or the development of their meta-cognitive abilities (Kuhn, 2000). Liu et al., (2017) describe conscious reflective activities in which learners conceptualize their knowledge and experiences through reconsideration and evaluation of their efforts for better outcomes or improvement (Öztürk and Karakuş, 2023). In the context of the relevant use of meta-cognitive skills and strategies in mathematics problem-solving among rural school learners, this study aims to investigate the impact of learners' meta-cognitive application on mathematics problem-solving.

Cognitive Developmental Theory

Cognitive Developmental Theory is a perspective on human development that highlights the significance of cognitive processes, including perception, attention, memory, and problem-solving, in shaping the course of development. It was initially proposed by Jean Piaget, a Swiss psychologist, who argued that children progress cognitively through a series of stages that build upon one another.

Piaget's Theory Outlines Four Stages of Cognitive Development

- *Sensorimotor stage* (birth to 2 years): During this stage, infants acquire knowledge about the world through their senses and motor actions. They develop a basic understanding of object permanence and causality.
- *Preoperational stage* (2 to 7 years): Children in this stage develop symbolic thinking and language skills. They can mentally represent objects and events but have limitations in comprehending abstract concepts.
- *Concrete operational stage* (7 to 12 years): Children in this stage develop logical thinking and can perform mental operations on concrete objects and events.
- *Formal operational stage* (12 years and beyond): Individuals in this stage develop abstract reasoning abilities and can think logically and systematically about abstract concepts.

According to Piaget, cognitive development is influenced by both biological maturation and environmental experiences. Children actively construct their knowledge through interactions with their environment, leading to the development of increasingly complex cognitive structures. Critics of Piaget's theory have suggested that it may overemphasize the role of biological maturation in cognitive development while underestimating the impact of social and cultural factors. Nevertheless, the theory remains a highly influential framework for understanding the cognitive development of children and adolescents.

Critical Thinking

Critical thinking is another essential approach to learning, which involves the ability to analyze and evaluate information, arguments, and evidence. It encourages students to question assumptions, consider multiple perspectives, and make informed decisions based on evidence. By developing critical thinking skills, students can better understand complex issues, communicate more effectively, and become more independent learners (Kumari et al., 2022). However, critical thinking can be challenging to teach and assess, as it involves a range of cognitive and affective skills. Critical thinking is more important now than ever before. In today's world, we are inundated with information from various sources, and it is increasingly difficult to discern fact from fiction. Critical thinking enables us to evaluate information logically and objectively and make informed decisions based on evidence and reason. With the rise of social media and the internet, misinformation and propaganda are rampant, and critical thinking skills are essential to avoid falling prey to false information. Critical thinking also enables us to evaluate different perspectives and opinions and engage in productive dialogue with others who may have different viewpoints.

In the workplace, critical thinking is a highly valued skill. Employers want employees who can analyze data, solve problems, and make informed decisions. Critical thinking also enables employees to innovate and adapt to changing circumstances, which is increasingly important in today's fast-paced and rapidly changing world. Overall, critical thinking is a vital skill in present times. It enables us to navigate the complex world around us, make informed decisions, and contribute to our communities and workplaces in a meaningful way. Critical thinking is an essential component of learning as it involves analyzing and evaluating information logically and objectively, and applying reasoning and problem-solving skills to make informed decisions. By incorporating critical thinking into the learning process, learners can deepen their understanding of concepts and develop higher-order thinking skills (Palupi et al., 2020).

When learners engage in critical thinking, they are actively processing and analyzing information, rather than simply memorizing it. This type of active engagement enhances their understanding of the material and enables them to apply what they have learned to new situations. Critical thinking also enables learners to evaluate sources of information and identify biases, which is particularly important in today's world of information overload. Incorporating critical thinking into learning can also help learners become more independent and self-directed (Jindal et al., 2016). By developing their questions and hypotheses, learners can take ownership of their learning and become more engaged in the process. Critical thinking also enables learners to identify gaps in their understanding and seek out additional information or resources to fill those gaps.

Overall, critical thinking is an essential component of learning. By encouraging learners to analyze and evaluate information logically and objectively, we can help them develop higher-order thinking skills, deepen their understanding of concepts, and become more independent and self-directed learners.

Historical Review of Collaborative Learning and Cooperative Learning

Collaborative learning and cooperative learning have a rich history in education, with roots dating back to the early 20th century. Here is a brief historical review of these two approaches:

Collaborative learning can be traced back to the 1920s, when John Dewey, a philosopher and educational reformer, promoted the idea that education should be centered around the needs and interests of the student, rather than the teacher (Dehghani et al., 2023). He advocated for a more student-centered approach that emphasized collaboration and inquiry-based learning. In the 1970s, a group of educators, including Elliot Aronson and David Johnson, began to develop a more formalized approach to collaborative learning, emphasizing the importance of group work and peer support in the learning process.

Cooperative learning can be traced back to the work of Lev Vygotsky, a Soviet psychologist, who believed that learning is a social process and that students can benefit from working together in groups. In the 1960s, a group of psychologists, including Robert Slavin and Roger Johnson, began to develop a more formalized approach to cooperative learning, emphasizing the importance of positive interdependence, individual accountability, and group processing in the learning process.

Over the years, both collaborative learning and cooperative learning have gained widespread acceptance in education, with many educators recognizing the benefits of these approaches in promoting student engagement, fostering critical thinking, and developing essential 21st-century skills such as communication, collaboration, and problem-solving. Today, these approaches continue to evolve and adapt to new technologies and changing educational contexts, and they remain important tools for promoting equitable and effective learning for all students (Howells, 2018).

Modern Trends in Collaborative Learning and Cooperative Learning

Collaborative learning and cooperative learning continue to evolve and adapt to new educational contexts and technologies. Here are some modern trends in these approaches:

- *Technology integration:* Technology is increasingly being used to support collaborative and cooperative learning. Online collaboration tools, such as Google Docs and Microsoft Teams, allow students to work together on projects and assignments in real-time, regardless of their physical location. Digital whiteboards and online brainstorming tools can also facilitate group work and brainstorming sessions.
- *Student-led learning*: There is a growing trend towards student-led learning, which involves giving students more autonomy and responsibility in their learning. This can be facilitated through collaborative and cooperative learning approaches, which encourage students to take ownership of their learning and work together to achieve common goals.
- *Flipped classroom model*: In a flipped classroom model, students are responsible for learning content outside of class, through online lectures, videos, or other resources. In-class time is then used for collaborative and cooperative activities, such as group projects or problem-solving exercises. This approach can help to maximize the benefits of collaborative and cooperative learning, by providing students with more time and opportunities to work together.
- *Diversity and inclusion*: There is a growing recognition of the importance of diversity and inclusion in education, and collaborative and cooperative learning approaches can play an important role in promoting these values. By working together in diverse groups, students can learn to appreciate and respect different perspectives and experiences and develop essential interpersonal skills.
- *Project-based learning:* Project-based learning involves students working together on a long-term, complex project that requires collaboration, problem-solving, and critical thinking skills. This approach can be facilitated through collaborative and cooperative learning approaches and can help to prepare students for real-world challenges in a variety of fields.

In summary, modern trends in collaborative and cooperative learning include technology integration, student-led learning, the flipped classroom model, a focus on diversity and inclusion, and project-based learning. These approaches continue to evolve and adapt to new educational contexts and challenges, and remain important tools for promoting equitable and effective learning for all students (de Oliveira and Mahtari, 2022).

Relationship between Collaborative and Cooperative Learning

Collaborative and cooperative learning are two related approaches to learning that emphasize the importance of social interaction and communication in the learning process. Collaborative learning involves students working together on a task or project, while cooperative learning involves students working together to achieve a common goal. Both approaches can help students develop important skills such as teamwork, communication, and problem-solving, as well as enhance their motivation and engagement in the learning process. However, effective collaboration and cooperation require careful planning, management, and assessment to ensure that all students are actively engaged and learning.

Collaborative and cooperative learning are important approaches to learning that enable learners to work together towards a common goal. These approaches foster teamwork, communication, problem-solving, and other skills that are essential for success in today's world. Collaborative learning involves learners working together to achieve a common goal. This approach enables learners to share ideas, knowledge, and resources, and to benefit from each other's strengths and perspectives. Collaborative learning can take many forms, such as group projects, peer reviews, or classroom discussions. Cooperative learning involves learners working together in a structured way to achieve a common goal. This approach typically involves assigning roles and responsibilities to each learner, and ensuring that each member of the group contributes to the group's success. Cooperative learning can be particularly effective in helping learners develop social and emotional skills, such as empathy, trust, and conflict resolution.

Collaborative and Cooperative Learning Are Important for Several Reasons

First, they enable learners to develop teamwork and communication skills, which are essential for success in the workplace.

Second, they enable learners to benefit from each other's strengths and perspectives, which can enhance their understanding and retention of information.

Third, they foster a sense of community and belonging, which can improve learners' motivation and engagement in the learning process.

Overall, collaborative and cooperative learning are important approaches to learning that enable learners to develop essential skills, deepen their understanding of concepts, and become more engaged and motivated learners.

Before World War II, social theorists such as Allport, Watson, Shaw, and Mead advocated for cooperative learning, suggesting that working in groups is more effective and efficient than working alone. However, it was in 1937 that researchers Mark Arthur May and Leonard William Doob collected the principles behind the Cooperative and Collaborative Learning Theory.

Theory of Cooperative and Collaborative Learning

Cooperative and Collaborative Learning refers to a learning situation where two or more individuals work together to have a shared educational experience. In these environments, students have the opportunity to learn from each other, utilize each other's skills and resources, and share experiences that benefit the entire group. The theory suggests that learners who engage in group settings and cooperate to achieve common goals are more likely to be successful, while those who work independently may exhibit competitive behaviors.

Cooperative and Collaborative Learning not only benefits academic development but also enhances learners' social skills and personalities by emphasizing teamwork and team spirit. In a classroom setting, this translates to higher levels of achievement, potential cross-ethnic friendships, lifelong interaction, improved communication and cognitive skills, and critical thinking.

The theory is based on five basic principles:

- *Interpersonal and collaborative skills*: Students learn how to work together and support each other. Instructors should encourage activities such as brainstorming, reflection, and active participation.
- *Face-to-face interaction:* Direct interaction among students makes the learning process dynamic. Students discuss their ideas, summarize them orally, and understand the value of individual differences and critical thinking.
- *Beneficial interdependence:* Students learn the importance of collaboration for successfully completing tasks, as well as the value of assigning team roles and effective representation.

- *Individual responsibility:* Despite collaboration, students should feel a sense of responsibility towards the group and understand the importance of their contribution to task completion. Slacking, relying on others' work, and merely following instructions are not conducive to success.
- *Group interaction processing:* Groups should learn how to interact effectively and then evaluate their level of collaboration and skills. Instructors should provide time for students to reflect on the group's dynamics, including participation, support, and smooth functioning.

STEM Education

STEM stands for Science, Technology, Engineering, and Mathematics. It is a term used to describe an interdisciplinary approach to education that integrates these four fields of study. STEM education aims to prepare students for the workforce of the future, as jobs in these fields are expected to grow at a much faster rate than other occupations. STEM education can take many forms, including traditional classroom instruction, hands-on projects, and experiential learning opportunities such as internships and apprenticeships. The goal of STEM education is to provide students with a strong foundation in science, technology, engineering, and mathematics, and to develop the critical thinking, problem-solving, and collaboration skills necessary for success in the 21st century.

STEM education is important because it prepares students for high-paying jobs in growing fields such as computer science, biotechnology, and renewable energy. It also helps to address the growing demand for workers with STEM skills, which many employers report is a significant challenge. By promoting STEM education, educators and policymakers hope to create a more competitive workforce and foster innovation and economic growth.

STEM (Science, Technology, Engineering, and Mathematics) education has become an increasingly important focus in modern education, as it prepares students for the high-tech jobs of the future. Cooperative learning strategies are an effective approach to teaching STEM subjects, as they allow students to work together to solve complex problems, learn from one another, and develop essential skills such as communication, collaboration, and critical thinking.

Some Effective Cooperative Learning Strategies for STEM Education Include

- *Jigsaw Method*: In this strategy, students are divided into small groups, and each group is assigned a specific topic or concept to learn. Students then research and become experts in their topic and then come back to teach their findings to the larger group.
- *Problem-Based Learning*: This strategy involves presenting students with a real-world problem that requires STEM skills to solve. Students work in groups to research, design, and test a solution to the problem.
- *Think-Pair-Share*: In this strategy, students work together in pairs to discuss a problem or concept and then share their thoughts and ideas with the larger group.
- *Peer Teaching*: This strategy involves having students take turns teaching each other different STEM concepts or skills. This approach reinforces their understanding while allowing them to practice teaching and leadership skills.

Cooperative Learning Contributes to Equitable Education

Cooperative learning strategies promote student engagement, foster critical thinking, and help to develop essential 21st-century skills that are highly sought after by employers. By integrating these strategies into STEM education, teachers can create a more collaborative and engaging learning environment that prepares students for success in the modern workforce. Cooperative learning is an approach to education that emphasizes group work and collaboration among students. This approach has been shown to contribute to more equitable education by promoting a sense of community in the classroom, encouraging students to work together and learn from one another, and providing opportunities for students of different backgrounds and abilities to succeed. One of the main benefits of cooperative learning is that it promotes a sense of community and shared responsibility in the classroom. When students work together in groups, they develop a sense of accountability to one another, and they are more likely to help each other succeed. This can be especially important for students from disadvantaged backgrounds, who may be more likely to feel isolated or disconnected from the classroom.

Cooperative learning also provides opportunities for students of different backgrounds and abilities to work together and learn from one another. By working in diverse groups, students can develop a greater understanding and appreciation for different perspectives and experiences, and they can learn to work collaboratively with people who may have different strengths and weaknesses.

Furthermore, cooperative learning can help to address some of the inequities that exist in traditional classroom settings. In many classrooms, students who are more outspoken or assertive may dominate discussions and group work, leaving other students with less opportunity to participate. Cooperative learning structures can provide more equitable opportunities for all students to contribute and participate in the learning process.

In summary, cooperative learning can contribute to more equitable education by promoting a sense of community, providing opportunities for diverse students to work together and learn from one another, and addressing some of the inequities that exist in traditional classroom settings.

There Are Several Models of Cooperative Learning, Including in STEM Education

- *Structured team learning:* This model involves the use of heterogeneous teams of four students based on their performance level, ability, sex, ethnicity, and socioeconomic status. The process consists of whole-class presentations, group discussions, individual tests, and group recognition based on students' improved scores.
- *Student teams-achievement division (STAD):* STAD is a model that can be used in various subjects such as mathematics, language, arts, and social studies. It follows a four-step process: whole-class presentation, group discussion, individual test, and group recognition based on student's performance compared to their past averages.
- *Teams-games-tournament (TGT):* TGT incorporates teacher presentations, teamwork, and group discussions similar to STAD. However, instead of individual tests, weekly tournaments are conducted. Group recognition is based on the results of these tournaments rather than improvement scores.
- *Cooperative integrated reading and composition (CIRC):* CIRC, developed by Slavin, focuses on specific materials for teaching

> reading and writing in upper elementary grades. It involves three major language activities: basal-related activities, direct instruction in reading comprehension, and integrated language arts/writing. Students are assigned to teams composed of two pairs from different ability groups, ensuring heterogeneity within each team.

In CIRC, each pair within the team has similar abilities, while the overall team comprises pairs with different abilities. Reading materials are assigned based on each pair's ability, providing equal opportunities for success among all team members.

Skill-Based Learning

Skill-based learning is an approach that focuses on developing specific skills or competencies, rather than just acquiring knowledge. This approach is particularly relevant in vocational and professional education, where students need to develop practical skills such as problem-solving, communication, and decision-making. Skill-based learning can be more engaging and motivating for students, as it emphasizes the practical application of knowledge and helps them develop skills that are directly relevant to their future careers. However, skill-based learning can also be challenging to assess, as it often involves complex, real-world tasks that are difficult to evaluate using traditional assessment methods.

In conclusion, meta-cognitive, critical thinking, collaborative, cooperative, and skill-based learning are all important approaches to learning that can help students develop the necessary knowledge, skills, and attitudes to succeed in their academic and professional lives. These approaches are not mutually exclusive and can be combined in various ways to create a rich and engaging learning experience. However, implementing these approaches effectively requires careful planning, management, and assessment, as well as ongoing support for both students and teachers. By embracing these approaches to learning, educators can help students become more independent, motivated, and successful learners.

Skill-based learning is becoming increasingly important in present times. With the rapid pace of technological change and the shifting demands of the global economy, workers must continually update and acquire new skills to remain competitive. Skill-based learning enables learners to acquire practical, job-relevant skills that can be immediately applied in the workplace. By focusing on the development of specific skills, learners can become more efficient, effective, and productive in their work. Skill-based learning can take many forms, such as online courses, workshops, apprenticeships, or on-the-job training.

In addition to benefiting individuals, skill-based learning is also important for businesses and organizations. By investing in the skills of their employees, businesses can improve their bottom line, increase productivity, and gain a competitive edge. Skill-based learning can also help businesses adapt to changing market demands and technologies, and ensure they have a skilled workforce to meet those demands. Skill-based learning is also important for society as a whole. As the economy becomes more knowledge-based, it is increasingly important to ensure that all members of society have the skills necessary to succeed in the workforce. By investing in skill-based learning, we can ensure that everyone has access to the training and education they need to succeed in today's rapidly changing world. STEM (Science, Technology, Engineering, and Mathematics) education and skill-based learning are both approaches to education that emphasize the development of practical skills and real-world problem-solving abilities.

STEM education focuses on teaching students about the interconnectedness of the four fields of study and how they can be applied to solve real-world problems. This approach emphasizes hands-on, experiential learning and encourages students to engage in critical thinking, problem-solving, and collaboration. STEM education aims to prepare students for careers in high-growth fields such as computer science, biotechnology, and renewable energy.

Skill-based learning, on the other hand, is an approach to education that focuses on the development of practical skills that are directly applicable to the workplace. This approach emphasizes hands-on training and experiential learning opportunities, and it often involves apprenticeships, internships, and other forms of work-based learning. Skill-based learning aims to prepare students for specific careers or industries by providing them with the practical skills and knowledge they need to succeed.

Both STEM education and skill-based learning are important approaches to education that can prepare students for success in the 21st-century workforce. By emphasizing practical skills and real-world problem-solving abilities, these approaches can help to bridge the gap between education and employment and create a more competitive and innovative workforce.

Overall, skill-based learning is essential in present times. By enabling learners to acquire practical, job-relevant skills, we can improve the competitiveness of individuals, businesses, and society as a whole, and ensure that we are prepared for the challenges and opportunities of the 21st century.

Strategies for Incorporating a Comprehensive and Multi-Faceted Approach Used in STEM Education

- *Curriculum Development*: Curriculum developers should focus on creating learning materials and activities that promote meta-cognitive, critical thinking, collaborative, cooperative, and skill-based learning. This can be done by incorporating real-world scenarios and problems, encouraging teamwork and peer learning, and using technology to enhance learning experiences.
- *Teacher Training*: Teachers should be trained in pedagogical approaches that support meta-cognitive, critical thinking, collaborative, cooperative, and skill-based learning. They should also be trained in how to use technology to support these approaches, as well as how to assess and provide feedback on these skills.
- *Classroom Management*: Classroom management should be structured to encourage meta-cognitive, critical thinking, collaborative, cooperative, and skill-based learning. This can be done by creating a supportive and inclusive learning environment, allowing for student-led discussions, and providing opportunities for group work and peer learning.
- *Assessment and Evaluation*: Assessments should be designed to measure students' meta-cognitive, critical thinking, collaborative, cooperative, and skill-based learning. This can be done through project-based assessments, portfolios, and rubrics that assess these skills. Teachers should also provide timely feedback to students to help them improve their skills.
- *Partnerships and Collaboration*: Schools should develop partnerships and collaborations with businesses, non-profits, and other organizations to provide students with opportunities for real-world learning experiences. This can include internships, apprenticeships, and mentorship programs.

Conclusion

Skill is the ability to perform a task effectively and efficiently, distinguishing between intellectual, motor, personal, and interpersonal skills. Skills are different from knowledge, as they are developed through experience and practice. The use of educational games can enhance engagement and provide tailored feedback to learners. Designing games with different levels of difficulty and incorporating various learning styles can accommodate learner preferences. The concept of transfer of training is also crucial, and game designers should provide similar problem-solving contexts to facilitate the transfer of skills.

Overall, incorporating meta-cognitive, critical thinking, collaborative, cooperative, and skill-based learning into the education system requires a comprehensive and multi-faceted approach that involves curriculum development, teacher training, classroom management, assessment and evaluation, and partnerships and collaborations. By implementing these strategies, we can prepare students for success in the 21st century.

References

Bakar, M. A. A., & Ismail, N. (2020). Express students' problem solving skills from metacognitive skills perspective on effective mathematics learning. *Universal Journal of Educational Research*, *8*(4), 1404-1412.

Chauhan, S., Gupta, P., Palvia, S., & Jaiswal, M. (2021). Information technology transforming higher education: A meta-analytic review. *Journal of Information Technology Case and Application Research*, *23*(1), 3-35.

Dehghani, E., Rezvani, E., & Hadian, B. (2023). The role of collaborative inquiry in Iranian EFL teachers' reflective teaching: the case of teaching experience. *Reflective Practice*, 1-16.

de Oliveira Biazus, M., & Mahtari, S. (2022). The impact of project-based learning (PjBL) model on secondary students' creative thinking skills. *International Journal of Essential Competencies in Education*, *1*(1), 38-48.

Ekici, M. (2021). A systematic review of the use of gamification in flipped learning. *Education and Information Technologies*, *26*(3), 3327-3346.

Facione, P. A. (2011). Critical thinking: What it is and why it counts. *Insight assessment*, *1*(1), 1-23.

Howells, K. (2018). The future of education and skills: education 2030: the future we want. Paris OECD.

Ismail, N. S., Harun, J., & Salleh, S. M. (2018). The Development and Students' Perception on Mobile Problem Based Learning (PBL) Science Dictionary Application. *Advanced Science Letters*, *24*(1), 6-8.

Johnson, D. W., & Johnson, R. T. (1987). *Learning together and alone: Cooperative, competitive, and individualistic learning*. Prentice-Hall, Inc.

Jindal, P., & Tej, F. (2016). Being an educator in private vs. public school system: A comparative study on job contentment. *IRA International Journal of Education and Multidisciplinary Studies*, *3*(3), 309-319.

Jain, S., Bajaj, B., & Singh, A. (2023). Linking Metacognition, Workplace Cognitive Competencies and Performance: An Integrative Review-Based Conceptual Framework. *Primenjena psihologija*, *16*(1), 107-147.

Kumari, S., Jindal, P., & Chand, P. K. (2022). Social Media Networking: As Facilitator of Acculturation. *ECS Transactions*, *107*(1), 9381.

Kuhn, D. (2000). Metacognitive development. *Current directions in psychological science*, *9*(5), 178-181.

Liu, C., Chen, S., Chi, C., Chien, K. P., Liu, Y., & Chou, T. L. (2017). The effects of clickers with different teaching strategies. *Journal of Educational Computing Research*, *55*(5), 603-628.

Öztürk, Ö. Y., & Karakuş, M. (2023). Examining Learning Styles, Creative Thinking Skills, and Academic Success of Eighth-Grade Students in Middle School. *Educational Academic Research*, (48), 42-58.

Perry, E., Owen, D., Booth, J., & Bower, K. (2019). The curriculum for initial teacher education: Literature review.

Palupi, B., Subiyantoro, S., Triyanto, T., & Rukayah, R. (2020). Creative-thinking skills in explanatory writing skills viewed from learning behaviour: A mixed method case study. *International Journal of Emerging Technologies in Learning (IJET)*, *15*(1), 200-212.

Roschelle, J., & Teasley, S. D. (1995). The construction of shared knowledge in collaborative problem solving. In *Computer supported collaborative learning*, 69-97. Springer Berlin Heidelberg.

Wiggins, G. (2012). Seven keys to effective feedback. *Feedback*, *70*(1), 10-16.

Chapter 8

Exploring the Effectiveness of Coding Education Methods and Their Impact on the Data Science Job Market Requirements: A Systematic Review

Nipuna Sankalpa Thalpage[1]
and Rohit Bansal[2,*], PhD
[1]Cardiff School of Technologies, Cardiff Metropolitan University, Cardiff, Wales, UK
[2]Department of Management Studies, Vaish College of Engineering, Rohtak, India

Abstract

The growing demand for data science and coding skills has led to an increase in the number of educational programs in these areas. However, the effectiveness of these programs in terms of career development outcomes remains unclear. This systematic review aims to evaluate the available literature on the impact of coding education on data science career development. The study will follow a rigorous and systematic approach to identify and analyze relevant research on data science and coding education. A comprehensive search of relevant databases will be conducted using keywords and MeSH terms related to data science, coding education, career development, and program evaluation. The purpose of this systematic review is to evaluate the available literature on the effectiveness of coding education for Data science career development. The increasing demand for professionals with data science and coding skills has led to a growth in the number of educational programs in these areas. However, the quality and effectiveness of these

* Corresponding Author's Email: rohitbansal.mba@gmail.com.

In: STEM
Editors: Priya Jindal, Amandeep Singh and Rashmi Aggarwal
ISBN: 979-8-89113-305-1

programs are not well understood. This systematic review aims to address this gap by synthesizing the existing research on the impact of coding education on data science career development.

Keywords: systematic review, coding education, data science, career development

Introduction

Data science has become increasingly important in various industries due to the growing demand for professionals with these skills. The field of data science, in particular, has been identified as a key driver of innovation and growth in many sectors (Davenport and Patil, 2012). As data becomes more prevalent in business and society, the ability to analyze and interpret it is essential for success in many career paths (Manyika et al., 2011).

The demand for individuals with coding skills has led to a proliferation of educational programs in these areas. In recent years, universities and other institutions have created a variety of programs aimed at teaching these skills, ranging from coding boot camps to full-time degree programs. However, the quality and effectiveness of these programs are not well understood, and there is a need for research to evaluate their impact on career development.

Importance of Coding Education for Data Science

Coding education is increasingly important for individuals seeking to develop their skills and advance their careers. The job market for professionals with these skills is growing rapidly, with employers across a variety of industries seeking candidates with expertise in data science. Furthermore, individuals with data science and coding skills tend to have higher salaries and greater job security than those without these skills (Karaoglu et al., 2022).

In addition to the potential benefits for individuals, data science skills are also important for the overall success of organizations. The ability to analyze and interpret data can provide insights that drive innovation and growth, leading to a competitive advantage for businesses (Manyika et al., 2011). Therefore, investing in coding education for Data Science can be beneficial for both individuals and organizations.

Purpose of the Systematic Review

The purpose of this systematic review is to evaluate the available literature on the effectiveness of coding education for data science career development. Specifically, the review aims to answer the following research questions.

Research Questions

1. What are the factors and methods that contribute to the effectiveness of coding education programs?
2. What are the main skills and requirements of the Data Science job market?

Literature Review

The body of research on coding education for data science career development is extensive and covers a wide range of disciplines. Miller (2019) focused on the promotion of mathematical knowledge and thinking, with a specific focus on the identification of mathematical patterns and structures, through engagement with coding activities. Feijóo-García et al., (2020) evaluated the performance of adult participants with no prior coding experience on conventional block-based programming platforms. Several studies have contributed to our understanding of key aspects of the field of coding education and related areas. Ari et al., (2022) conducted research on early childhood preservice teachers' perceptions of computer science and gender stereotypes. Their findings shed light on the importance of addressing gender biases and stereotypes to promote inclusive computer science education. In line with this, the diverse range of research on data science and coding education highlights the significance of this topic in modern-day education and the need for further investigation into its effectiveness in career development.

Fischer et al., (2021) explored the impact of coding platforms such as Codewars and CodinGame on undergraduate students' motivation in coding education. Their study revealed a positive effect on student motivation, emphasizing the potential of such platforms to enhance learning experiences. Hundhausen et al., (2013) presented a set of best practices for implementing

Pedagogical Code Reviews (PCR) in software engineering education. Their work provides valuable insights for educators teaching PCRs, offering guidance on effective teaching practices in this area.

Prato (2017) explores the use of programming logic and coding languages as educational tools for young children, elementary and middle school children, and adults. The study emphasizes the importance of providing access to coding education at an early age to foster technological literacy and promote career opportunities in the tech industry. Duda et al., (2021) aimed to cover basic coding topics and data science concepts through live coding and hands-on practice. Emery et al., (2021) provide an important window into how data science is taught in higher education biology programs and how instructors across disciplines can best move forward to empower their students.

Another study conducted by Karahan et al., (2015) suggested that media design processes could serve as an effective instructional approach in STEM education. The research findings indicated positive impacts on students' attitudes toward science and technology classes.

The growing importance of data science has captured considerable interest in various industries, leading organizations to prioritize the establishment of their data science capabilities. Nonetheless, practitioners often face obstacles when it comes to creating technology-driven activities based on data (Kumari et al., 2022). In a related context, Pournaras (2017) sheds light on the insights gained from designing and instructing a unique data science course that encompasses multiple disciplines. The study emphasizes the significance of collaboration among stakeholders and the implementation of a comprehensive curriculum to address these challenges effectively. By drawing attention to the practical experiences shared by (Pournaras, 2017) and the difficulties identified by (Kumari et al., 2022), a clearer connection emerges, underscoring the importance of addressing the barriers in technology-driven data science activities through collaborative efforts and robust educational frameworks.

Mikroyannidis et al., (2018) propose a system for enhancing data science accreditation using blockchain technology. In contrast, Oesterreich et al., (2019) argue that there is no evidence to suggest that data science skills are essential. Instead, they study the impact of digitization on the controller's job roles, skills, and competencies.

Mikalef et al., (2019) propose an explanatory model to understand the gap between intentions and actions regarding data science. They build their model on survey responses from key executives in Norwegian firms. Horton and

Hardin (2021) discuss ways to teach computational thinking in the special issue of the Journal of Statistics and Data Science Education.

This literature review provides evidence for the effectiveness of data science and coding education in preparing individuals for career development. The literature suggests that traditional data science programs and online platforms such as Codecademy and Udemy have been effective in increasing coding skills and career opportunities. Additionally, the study of Dart and Flutter has gained popularity and is effective in increasing job opportunities and salaries. However, the review also highlights the challenges and barriers to the successful implementation of coding education programs for the data science job Market. These challenges include a lack of access to resources, difficulty in finding qualified instructors, and a rapidly changing industry landscape that requires continuous adaptation of curriculum and teaching methods. To address these challenges, best practices have been identified for the effective implementation of data science and coding education programs, including incorporating practical projects and real-world applications, promoting collaboration and peer learning, and continuous evaluation and improvement of the curriculum.

Further, considering the pieces of literature on coding education for data science career development, there is still a gap in exploring the effectiveness of coding education methods and their impact on the data science job market requirements. While the literature has covered a wide range of disciplines and topics, including laboratory tasks, argumentation, information systems curricula, and interdisciplinary approaches, there is a need for further investigation into best practices for implementing coding education in data science. Several studies have proposed different methods, such as live coding and hands-on practice, media design processes, web scraping, and natural language processing, but there is a lack of research on the outcomes and impact of these approaches on the data science job market requirements. Understanding the effectiveness of coding education methods in preparing individuals for the data science job market is crucial for educators, employers, and job seekers alike. Finally, these kinds of literature also underscore the importance of coding education in preparing individuals for data science and careers in the current job market. As technology continues to advance and data-driven decision-making becomes increasingly important across industries, educational institutions, and training programs must develop and implement effective programs that meet the needs of the industry.

Methodology

To conduct this systematic review, we followed a streamlined approach that involved four stages:

1. Identifying relevant studies
2. Screening the studies based on eligibility criteria
3. Extracting data from the selected studies
4. Synthesizing the results.

First, the author conducted a comprehensive search of two databases: Google Scholar and IEEE Explore which are relevant to the field of data science education. We used a combination of keywords and MeSH terms such as "data science education," "coding education," "programming education," "data literacy," "data skills," "curriculum development," "industry needs," and "job market." The search was limited to articles published in English between 2010 and 2022 that focused on data science and coding education.

Secondly, we screened the articles based on our inclusion and exclusion criteria.

The inclusion criteria were:

1. Studies focused on data science and/or coding education
2. Studies examined the effectiveness of educational programs or curriculum development
3. Studies reported on the skills and knowledge required for data science
4. Studies that are published in a peer-reviewed journal.

The exclusion criteria were

1. Studies that focused on other subject areas than data science or coding education.
2. Studies that were not focused on education, curriculum development, or data science,
3. Studies that were not peer-reviewed,
4. Studies that were published before 2010.

Third, we used thematic analysis, which is a technique used to recognize, examine, and interpret patterns or themes that convey significance within qualitative data (Braun and Clarke, 2016) to analyze the selected studies. We organized the data into themes and patterns based on the research questions and objectives. Themes were developed through a process of coding and categorization of the data, with a focus on identifying patterns and relationships in the data. We used both inductive and deductive approaches to ensure that the themes were grounded in the data but also aligned with our research objectives.

Lastly, we synthesized the results of the selected studies and presented them in a narrative synthesis. We identified common themes and patterns across the studies and summarized them. By using thematic analysis, we aimed to provide a comprehensive and meaningful interpretation of the data that can inform future research and practice in the field of data science and coding education.

Although we have used PRISMA guidelines, A method that involves a well-defined and unambiguous research question and utilizes explicit and systematic methods to identify, select, and assess pertinent research which also involves collecting and analyzing data from the selected studies critically (Moher et al., 2009). Certain steps had to be skipped considering the duration and other limitations we had to deal with.

Overall, our approach aimed to provide a rigorous and systematic review of the literature, while also ensuring the feasibility of completing the study within a limited timeframe.

Data Collection

To identify relevant papers on data science and coding education, we conducted a search using two databases: Google Scholar and IEEE. The search was limited to papers published in English between 2010 and 2022. We used a combination of keyword searchers and operators, including "data science education," "coding education," "programming education," "data literacy," "data skills," "curriculum development," "industry needs," and "job market.

Duplicate papers were removed, and the remaining papers were assessed for relevance based on the research question and inclusion criteria. We identified a total of 16 papers from the screening process. Additionally, citation tracking was carried out in both the forward and backward directions to identify any additional relevant articles, which resulted in the inclusion of 2 papers in the final analysis (Greenhalgh and Peacock, 2005).

Using Google Scholar and IEEE to search for papers on data science and coding education allowed for a comprehensive and focused approach to identifying relevant papers. By using keyword searchers and operators, the search was able to target specific concepts and themes related to the research question. Furthermore, citation tracking helped to identify additional relevant papers and strengthen the study's findings (Booth, 2016).

Screening

After the search results were exported from the two databases, they were imported into a screening tool called Covidence which can significantly simplify the process of reviewing titles and abstracts for a JBI systematic review with an online platform that is available for free and also offers a subscription version that includes additional features (Macdonald et al., 2016). This tool allowed for a more efficient and streamlined approach to screening the search results.

The first step of the screening process involved reviewing the titles and abstracts of the records to determine their relevance to the research question and inclusion criteria. If the title and abstract indicated that the record may meet the inclusion criteria, the record was selected for full-text screening.

During the full-text screening process, the inclusion and exclusion criteria were applied more rigorously. Records were reviewed in depth to determine if they met the criteria for inclusion in the systematic review. If a record did not meet the criteria, it was excluded from further consideration.

Throughout the screening process, reasons for exclusion were recorded in a separate document. This allowed for transparency and traceability in the screening process, ensuring that decisions were made systematically and consistently.

Overall, the use of a screening tool and a transparent process for recording exclusion reasons helped to ensure that the final selection of studies was rigorous and unbiased. This approach increases the reliability and validity of the systematic review and enhances the quality of the findings. 18 papers were initially used for the screening process and after the screening 09 papers became eligible for the next phase of extracting data.

Data Extraction

In this systematic review, data extraction was performed on a total of 9 selected papers. The following sections were extracted from each paper: Study

ID, Study Design, Participants, Interventions/Exposures, Outcomes/Results, Limitations, and Other Notes.

Study ID

The unique identifier for each study was extracted, such as the author's name and publication year.

Study Design

The study design of each paper was extracted, such as randomized controlled trials, cohort studies, case-control studies, or cross-sectional studies.

Participants

The characteristics of the study participants were extracted, such as age, gender, ethnicity, and any relevant medical conditions.

Interventions/Exposures

The interventions or exposures used in each study were extracted, such as medication or treatment regimens, lifestyle modifications, or environmental exposures.

Outcomes/Results

The outcomes or results of each study were extracted, such as changes in disease incidence, symptom improvement, or adverse effects.

Limitations

Any limitations or potential biases in each study were extracted, such as small sample size, incomplete data, or confounding factors.

Based on the extracted data, a summary of the key findings and trends among the nine selected papers was compiled. For example, it may be noted that the majority of the studies were randomized controlled trials that focused on the effectiveness of coding knowledge for Data Science. Limitations may include small sample sizes or lack of diversity among study participants. Overall, the data extracted from each paper was used to inform the conclusions and recommendations presented in the systematic review. Below are the research papers that were used for data extraction along with the data that was extracted for the data coding process.

1. *Skill Requirements in Big Data: A Content Analysis of Job Advertisements* (Gardiner et al., 2016) in their research paper explores the content analysis of job advertisements. For this, the participants were online job ads related to big data. The study used the big data disciplinary conceptual model of professional expertise, abilities, and skills (BDDSCM). The paper highlighted that many big data personnel create analytical systems, and many big data positions demand traditional development abilities. The results of our research also showed that businesses still place a high priority on soft skills for big data personnel. The paper provides the future scope that the current study may have been skewed toward technical positions, underrepresenting big data-related positions that were more heavily weighted toward managerial decisions or simply statistical analysis.
2. *Integrating Online Meta-Cognitive Learning Strategy and Team Regulation to Develop Students' Programming Skills, Academic Motivation, and Refusal Self-Efficacy of Internet Use in a Cloud Classroom* (Tsai et al., 2022) the research paper used a Quasi-experimental design for the university students were the participants. The study used the team management and online metacognitive learning plan integration model. The study explores the Students' programming design abilities and self-efficacy with Internet use in a virtual classroom both greatly improved after taking an online programming course. The results of this study did not reveal the anticipated impacts of meta-cognitive learning strategy (MCLS) on students' programming abilities, academic motivation, or resistance to self-efficacy of Internet usage. A modest sample size was used for the investigation. The study solely looked at how combining team regulation (TR) and online meta-cognitive learning strategy (MCLS) affected students' development of programming abilities, academic motivation, and resistance to self-efficacy of Internet usage.
3. *Game-Based Learning: Enhancing Student Experience, Knowledge Gain, and Usability in Higher Education Programming Courses* (Zhao et al., 2022) in their current study used over 100 students from 3 institutions took part in the study. 78, 10, and 34 students took part in the Variable game and 65, 10, and 30 students participated in the Loop game in DCU, NCI, and STUBA, respectively. The study makes the case that game-based learning is a useful intervention to improve the student experience, knowledge acquisition, and usability in higher programming education. A comprehensive assessment

methodology which includes surveys, observations, and interviews was used. The pilot study's findings revealed that the majority of students, regardless of their geographic region, thought the games were both aesthetically pleasing and flawlessly functional. The majority of students enjoyed playing serious games because they believed that they improved their understanding of programming topics, piqued their interest in their classes, and helped them get higher grades. These outcomes show how effectively the suggested games accomplished their objectives. The study, which focuses on information retention, learner experience, and game usability, looked at how students responded to game-based learning. Based on their location, educational background, and the game they played, the study discovered that all pupils profited from the games, albeit to varied degrees. The study's limitations are mostly related to the technology that supports GBL, including the possibility of gadget addiction, poor posture, and decreased participant interest in other activities.

4. *Toward Knowledge Discovery Framework for Data Science Job Market in the United State* (Heidarysafa et al., 2022) the study highlighted that Python is the most popular language for work in data science, followed by R and Java. This shows that Python is not only the most popular language but also that, based on the trend and the gap, it would likely continue to be in demand. A data analyst primarily needs abilities for data retrieval (e.g., SQL, Excel, database), and visualization (e.g., Tableau, Power BI). Machine learning engineer job ads typically reflect a dominance of talents such as programming, machine learning, cloud, and big data technologies. On the other hand, an applicant for a position as a data scientist would require expertise in both of the aforementioned areas. Another significant finding from this study emphasizes the value of deep learning, cloud computing, and big data expertise for data scientists. Python appears more frequently than R, indicating that it is a more prevalent language for occupations using data in the industry than R, which is another interesting finding in this context. The study was based on Quantitative Analysis. Using the Knowledge Discovery Framework the research was limited only to the United States context.
5. *Problem-Solving by 5-6 Years Old Kindergarten Children in a Computer Programming Environment: A Case Study* (Fessakis et al., 2013) the study used the participant's 5-6-year-old kindergarten

children. The study was based on an exploratory case study. The study found that children enjoyed engaging in learning activities and had opportunities to develop mathematical concepts, problem-solving, and social skills. The intervention was designed as a part of the structured learning activities of the kindergarten which are teacher-guided and are conducted in a whole class social mode.

6. *School Perceptions of Coding Education in K-12: A Large Scale Quantitative Study to Inform Innovative Practice* (Wong et al., 2015) the study was conducted on 42 primary and secondary schools in Hong Kong. The data was collected through a questionnaire. The article presents the findings of a study conducted in 42 Hong Kong schools. The findings indicate that there is a movement in the neighborhood schools to promote computer science education, with favorable views of both teaching and learning. However, the study also mentions issues with education that still need to be resolved, such as a lack of teacher training and an inconsistent curriculum. The limitation of the study is that the results may not apply to other locations as the study only examined schools in Hong Kong.
7. *Teaching Introductory Programming: A Quantitative Evaluation of Different Approaches* (Koulouri et al., 2013) the study used pair programming, the Media Computation approach, and Peer Instruction. The data was collected through Quantitative Analysis. The participants were a module leader, 2 supporting lecturers, 6 graduate teaching assistants, and the students in the class. The paper makes the following claims: it is important to rigorously control the process in a naturalistic setting; some objective measurements should be taken into consideration to assess the effects of the strategies employed; quantitative evidence has been produced from these cycles of data collection and analysis; and recommendations for the design of CS1 curricula have been framed as a result of these analyses. The study indicated that even while students liked it and found it simple to use, programming in Scratch did not significantly affect their ability to solve problems. According to the study, assessment criteria for instructional interventions should be rephrased as "indicators of learning."
8. *The Effects of Teaching Programming via Scratch on Problem-Solving Skills: A Discussion from Learners' Perspective* (Kalelioglu and Gülbahar, 2014) the study was conducted on 49 primary school students. Explanatory sequential mixed methods design is used to

teach students from scratch. The study found that no significant increase in the mean of the factor of "self-confidence" in their problem-solving ability occurs.

9. *What is a Data Scientist? Analysis of Core Soft and Technical Competencies in Job Posting* (da Silveira et al., 2020) the study highlighted that Python, English, SQL, and good communication skills are the primary competencies needed for data science. According to Costa and Santos (2017), data scientists should have a profile that emphasizes their understanding of programming languages, databases, and machine learning. Google and Facebook, as quoted in Costa & Santos (2017) have also drawn attention to the importance of analytical, mathematical, and statistical skills as well as the passion for using data to make choices and address issues. All of those qualities are frequently cited in the positions under analysis, along with others. The fact that hard talents are listed in every opportunity, even if just briefly, indicates that the job postings value them the most. The data itself has a very dynamic nature since new jobs are being posted and fulfilled, which causes them to be deleted from the platform.

Data Synthesis

Data synthesis for this research was conducted using the Thematic Analysis process.

Firstly, the text from the outcomes and results section of the Data extraction was used to create codes by extracting significantly relevant texts to Research questions. These codes were specific phrases or sentences that were related to the research questions and provided insight into the data mentioned in Table 1.

Table 1. Data coding process

What are the factors and methods that contribute to the effectiveness of the data science and coding education program?	What are the main coding skills and requirements of the data science job market?
Online programming courses significantly improved students' programming design skills.	The top language for data science jobs is Python followed by R and Java.

Games helped them to understand better programming concepts;	The main skills required for data science are good communication, team player, problem-solving, Python, English, and SQL.
Students benefitted from the game, although to different degrees depending on their location, educational background, and the game played.	Hirers care more about the applicant's skills and knowledge than about his or her formal academic education.
Children enjoyed the engaging learning activities and had opportunities to develop mathematical concepts, problem-solving, and social skills.	Many big data employees develop analytical systems.

Source: Author's compilation.

Secondly, Table 2 was created by dividing the codes into two columns according to the research questions. This helped to organize the codes and ensured that each code was assigned to the appropriate research question.

Table 2. Organize codes relevant to research questions

Outcomes/ Results	Codes
Skills associated with many big data jobs require traditional development skills, reflecting that many big data employees develop analytical systems. Furthermore, the Analysis also highlighted that soft skills are still highly valued by employees for big data employees.	Many big data employees develop analytical systems, soft skills are still highly valued by employers for big data employees.
Online programming course significantly improves students' programming design skills and their refusal of self-efficacy of internet use in a cloud classroom. However, the expected effects of meta-cognitive learning strategy (MCLS) on developing students' programming skills, academic motivation, and refusal of self-efficacy of internet use were not found in this study.	Online programming courses significantly improved students' programming design skills.

Source: Author's compilation.

Thirdly, themes were created by connecting and combining similar codes. This step involved reviewing the codes and identifying commonalities and patterns in the data. The codes were then grouped into themes based on these commonalities and patterns.

The final output of the data analysis was a set of themes that provided insight into the research questions. These themes were supported by the codes and provided a deeper understanding of the data.

Themes Derived from the Codes

Theme 1: Kids Enjoy Learning Coding with Fun Activities and Aesthetics

This theme highlights the importance of incorporating fun activities like Gaming and aesthetics in coding education to make it more enjoyable for children. It emphasizes that coding should be taught in a way that is both informative and entertaining to capture children's attention and interest. By doing so, children can learn to code more effectively and improve their problem-solving, critical thinking, and social skills. The theme also emphasizes that coding education should be designed to appeal to different learning styles, allowing children to learn at their own pace and in a way that suits them best.

Theme 2: More Training and Revised Curriculum Required for Coding Education

This theme emphasizes the need for more teacher training and a revised curriculum to ensure the successful delivery of coding education. It highlights the challenges faced by educators in teaching programming, such as the lack of standardized curriculum and teacher training. The theme advocates for a more comprehensive and standardized curriculum that provides students with the necessary skills and knowledge to succeed in the field of programming. It also emphasizes the importance of providing teachers with adequate training and support to ensure that they are equipped to teach coding effectively.

Theme 3: Programming Education can Benefit Children by Improving their Personal Skills

This theme highlights the positive outcomes and benefits of programming education for children, such as an improvement in their personal skills. The theme emphasizes that coding education can help children develop problem-solving, critical thinking, creativity, and social skills. These skills are not only important for success in programming but also personal growth and development. The theme also highlights the potential for coding education to improve children's academic performance and self-confidence in their problem-solving abilities. Ultimately, programming education can provide children with valuable skills and knowledge that will benefit them in various aspects of their lives.

Theme 4: In the Field of Data Science Jobs, a Candidate's Soft Skills and Technical Abilities such as Coding Hold More Significance than their Academic Qualifications

This theme highlights the changing landscape of hiring practices in the field of data science, where soft skills and technical abilities hold more weight than traditional academic qualifications. It focuses on the top programming languages preferred by employers in the field of data science, which are Python, R, and Java, and the primary skills required for data scientists, such as good communication, problem-solving, team playing, and proficiency in programming languages like Python and SQL. The theme also delves into the shifting priorities of hirers who now prioritize a candidate's skills and knowledge over formal academic education. Additionally, it touches upon the fact that many big data employees develop analytical systems, and employers still value soft skills, like communication and teamwork, in addition to technical abilities.

Conclusion

The findings from the themes discussed above reveal the significance of incorporating fun and engaging activities in coding education to improve children's learning outcomes. Additionally, it is crucial to provide teachers with the necessary training and a standardized curriculum to ensure the successful delivery of coding education.

Moreover, programming education has the potential to enhance children's personal skills, such as problem-solving, critical thinking, and creativity, that will benefit them in various aspects of their lives. This implies that programming education can serve as a valuable tool for personal growth and development. In the field of data science jobs, the significance of soft skills and technical abilities, particularly coding, has surpassed traditional academic qualifications. Employers prioritize the skills and knowledge of candidates over formal academic education, and programming languages like Python, R, and Java have become essential for data scientists. Good communication, problem-solving, and team-playing skills are also vital for success in this field. Thus, to succeed in the rapidly evolving field of data science and programming, it is essential to focus on developing technical abilities, like coding, as well as soft skills. Overall, programming education offers a range of benefits and outcomes, making it an essential tool for personal and professional growth in today's world.

The findings on coding education and the field of data science are closely linked. As children are introduced to coding education at a young age, they can develop fundamental skills in problem-solving, critical thinking, and programming languages like Python, which are essential for data science jobs. The emphasis on incorporating fun activities and appealing to different learning styles in coding education can also foster creativity and communication skills, which are valuable assets in the data science job market. The need for more teacher training and a revised curriculum in coding education aligns with the demand for a comprehensive skill set in the field of data science, as employers prioritize candidates who possess both technical abilities and soft skills like communication and teamwork. Thus, investing in coding education can not only benefit children but also prepare them for a promising career path in the ever-growing field of data science.

This research was conducted under some limitations like time & number of reviewers available, therefore some steps had to skip and proceed in a rapid review manner. On a future research note, this can be done by expanding inclusion and exclusion criteria further by gathering papers from more wide time frame and also gathering papers from different contexts and languages. Also, the methodology used in this systematic review could be applied to various research questions in the field of data science and coding education, with potential future research directions focusing on specific target populations or the skills and knowledge required for data science and coding careers.

References

Ari, F., Arslan-Ari, I., & Vasconcelos, L. (2022). Early childhood preservice teachers' perceptions of computer science, gender stereotypes, and coding in early childhood education. *TechTrends*, *66*(3), 539-546.

Booth, A. (2016). Searching for qualitative research for inclusion in systematic reviews: a structured methodological review. *Systematic reviews*, *5*(1), 1-23.

Braun, V., & Clarke, V. (2016). (Mis) conceptualising themes, thematic analysis, and other problems with Fugard and Potts'(2015) sample-size tool for thematic analysis. *International Journal of social research methodology*, *19*(6), 739-743.

da Silveira, C. C., Marcolin, C. B., da Silva, M., & Domingos, J. C. (2020). What is a Data Scientist? Analysis of core soft and technical competencies in job postings. *Revista Inovação, Projetos e Tecnologias*, *8*(1), 25-39.

Davenport, T. H., & Patil, D. J. (2012). Data scientist. *Harvard business review*, *90*(5), 70-76.

Duda, M., Sovacool, K. L., Farzaneh, N., Nguyen, V. K., Haynes, S. E., Falk, H., & Wolford, B. N. (2021). Teaching Python for Data Science: Collaborative development of a modular & interactive curriculum. *The Journal of open source education*, *4*(46).

Emery, N. C., Crispo, E., Supp, S. R., Farrell, K. J., Kerkhoff, A. J., Bledsoe, E. K., & Aiello-Lammens, M. E. (2021). Data science in undergraduate life science education: a need for instructor skills training. *BioScience*, *71*(12), 1274-1287.

Feijóo-García, P. G., McNamara, K., & Stuart, J. (2020). The Effects of Native Language on Block-Based Programming Introduction: A Work in Progress with Hispanic Population. *Research on Equity and Sustained Participation in Engineering, Computing, and Technology (RESPECT)*, 1, 1-2. IEEE.

Fessakis, G., Gouli, E., & Mavroudi, E. (2013). Problem solving by 5–6 years old kindergarten children in a computer programming environment: A case study. *Computers & Education*, *63*, 87-97.

Fischer, M., Gerdtham, U. G., Heckley, G., Karlsson, M., Kjellsson, G., & Nilsson, T. (2021). Education and health: long-run effects of peers, tracking and years. *Economic Policy*, *36*(105), 3-49.

Gardiner, A., Aasheim, C., Rutner, P., & Williams, S. (2018). Skill requirements in big data: A content analysis of job advertisements. *Journal of Computer Information Systems*, *58*(4), 374-384.

Greenhalgh, T., & Peacock, R. (2005). Effectiveness and efficiency of search methods in systematic reviews of complex evidence: audit of primary sources. *Bmj*, *331*(7524), 1064-1065.

Heidarysafa, M., Kowsari, K., Bashiri, M., & Brown, D. E. (2022). Towards a Knowledge Discovery Framework for Data Science Job Market in the United States. In *Proceedings of the Future Technologies Conference (FTC) 2021, 1*, 875-887. Springer International Publishing.

Horton, N. J., & Hardin, J. S. (2021). Integrating computing in the statistics and data science curriculum: Creative structures, novel skills and habits, and ways to teach computational thinking. *Journal of Statistics and Data Science Education*, *29* (1), 1-8.

Hundhausen, C. D., Agrawal, A., & Agarwal, P. (2013). Talking about code: Integrating pedagogical code reviews into early computing courses. *ACM Transactions on Computing Education (TOCE)*, *13*(3), 1-28.

Jindal, P., & Tej, F. (2016). Being an educator in private vs. public school system: A comparative study on job contentment. *IRA International Journal of Education and Multidisciplinary Studies*, *3*(3), 309-319.

Kalelioglu, F., & Gülbahar, Y. (2014). The Effects of Teaching Programming via Scratch on Problem Solving Skills: A Discussion from Learners' Perspective. *Informatics in education*, *13*(1), 33-50.

Karahan, E., BİLİCİ, S. C., & Ayçin, Ü. N. A. L. (2015). Integration of media design processes in science, technology, engineering, and mathematics (STEM) education. *Eurasian Journal of Educational Research*, *15*(60), 221-240.

Karaoglu, G., Hargittai, E., & Nguyen, M. H. (2022). Inequality in online job searching in the age of social media. *Information, Communication & Society*, *25*(12), 1826-1844.

Koulouri, T., Lauria, S., & Macredie, R. D. (2014). Teaching introductory programming: A quantitative evaluation of different approaches. *ACM Transactions on Computing Education (TOCE)*, *14*(4), 1-28.

Kumari, S., Jindal, P., & Chand, P. K. (2022). Social Media Networking: As Facilitator of Acculturation. *ECS Transactions*, *107*(1), 9381.

Macdonald, M., Misener, R. M., Weeks, L., & Helwig, M. (2016). Covidence vs Excel for the title and abstract review stage of a systematic review. *JBI Evidence Implementation*, *14*(4), 200-201.

Manyika, J., Chui, M., Brown, B., Bughin, J., Dobbs, R., Roxburgh, C., & Hung Byers, A. (2011). *Big data: The next frontier for innovation, competition, and productivity*. McKinsey Global Institute.

Mikalef, P., Boura, M., Lekakos, G., & Krogstie, J. (2019). Big data analytics and firm performance: Findings from a mixed-method approach. *Journal of Business Research*, *98*, 261-276.

Mikroyannidis, A., Domingue, J., Bachler, M., & Quick, K. (2018). Smart blockchain badges for data science education. *IEEE Frontiers in Education Conference (FIE)*, 1-5. IEEE.

Miller, J. (2019). STEM education in the primary years to support mathematical thinking: Using coding to identify mathematical structures and patterns. *Zdm*, *51*(6), 915-927.

Moher, D., Liberati, A., Tetzlaff, J., Altman, D. G., & PRISMA Group*, T. (2009). Preferred reporting items for systematic reviews and meta-analyses: the PRISMA statement. *Annals of internal medicine*, *151*(4), 264-269.

Oesterreich, T. D., Teuteberg, F., Bensberg, F., & Buscher, G. (2019). The controlling profession in the digital age: Understanding the impact of digitisation on the controller's job roles, skills and competences. *International journal of accounting information systems*, *35*(C).

Pournaras, E. (2017). Cross-disciplinary higher education of data science–beyond the computer science student. *Data Science*, *1*(1-2), 101-117.

Prato, S. C. (2017). Beyond the computer age: A best practices intro for implementing library coding programs. *Children and Libraries*, *15*(1), 19-21.

Tsai, C. W., Lee, L. Y., Cheng, Y. P., Lin, C. H., Hung, M. L., & Lin, J. W. (2022). Integrating online meta-cognitive learning strategy and team regulation to develop students' programming skills, academic motivation, and refusal self-efficacy of Internet use in a cloud classroom. *Universal Access in the Information Society*, 1-16.

Wong, G. K., Cheung, H. Y., Ching, E. C., & Huen, J. M. (2015). School perceptions of coding education in K-12: A large scale quantitative study to inform innovative practices. *IEEE international conference on teaching, Assessment, and learning for engineering (TALE)*, 5-10. IEEE.

Zhao, D., Muntean, C. H., Chis, A. E., Rozinaj, G., & Muntean, G. M. (2022). Game-based learning: enhancing student experience, knowledge gain, and usability in higher education programming courses. *IEEE Transactions on Education*, *65*(4), 502-513.

Chapter 9

The Utilization of Science Parks, Museums, Zoos and Aquariums (SPMZAs) for Enriching STEM Education in Developing South Asian Countries in Today's Digital World

Joy Mukhopadhyay[1], PhD
Arka Ghosh[2,*]
and Ranajit Chatterjee[3], PhD
[1]T. John College, Bengaluru, Karnataka, India
[2]Corporate Strategy & Business Transformation, Leading Global Financial Institution, Bengaluru, Karnataka, India
[3]Department of Research and Development, University of Electro-Communications, Chofu, Kyoto, Japan

Abstract

The term STEM (Science, Technology, Engineering, and Mathematics) was coined not long back to emphasize that the study of the included four disciplines holds the key to technological and economic development. Accordingly, a very high number of institutions of higher learning in India offer a wide variety of STEM courses. The craze for joining these courses is evident from the humongous number of students opting for them, year after year. The biggest challenge in developing South Asian countries in conducting these courses is the lack of resources (financial, technical, and human). In most places, the financial crunch as well as lack of willingness from the providers as well as the provided, result in either poorly furnished laboratory facilities or in worst cases, no

---

[*]Corresponding Author's Email: arka829@gmail.com.

In: STEM
Editors: Priya Jindal, Amandeep Singh and Rashmi Aggarwal
ISBN: 979-8-89113-305-1

laboratories at all. This can be reduced to a great extent by collaborative efforts between educational institutions and public or private science parks, museums, zoos, and aquariums (SPMZAs). This chapter ideates on using the existing SPMZA infrastructure in various parts of South Asia and how educational institutions can collaborate with these SPMZA facilities to facilitate and enhance STEM education. This chapter is a great addition to the equal opportunity in education line of thought. We also tell how VR (Virtual Reality) and AR (Augmented Reality) adoption further upgrades our propositions, and from them, we present an adoptable model as well.

Keywords: STEM, collaboration, SPMZA, virtual reality, augmented reality

Introduction

In this chapter, we have proposed a model to enrich STEM education in developing South Asian economies. In today's time, when equality in education on all fronts is becoming a global voice, this chapter would be a great contribution to that voice and the pertaining academic conversation. We start off the chapter by understanding what is STEM and its importance. Then we look at the pedagogy of teaching STEM in general, and we present the case of STEM education in Japan, which has a very advanced and holistic system of dispersion of STEM education. The next section then talks about the importance of SPMZAs in experiential learning, and then we present our propositions. While conceptualizing the propositions, we realized that individually they fall short of presenting our idea. Thus, we take a step further and develop a conceptual model, falling back on our propositions. Even though the model is conceptual, the tenets it relies upon are very real and as a matter of fact, present in our lives to date, thereby making this model adaptable in the real world. The best practices to integrate STEM in STEM education or otherwise are not yet put forth (English, 2016; Kelley and Knowles, 2016; Herschbach, 2011). This chapter is a humble and praxis attempt in this direction, in our South Asian context.

STEM & Its Importance

STEM, unabbreviated as science, technology, engineering, and mathematics – the acronym stands for the field and curriculum centered around the disciplines individually, as well as in any permutation and combination amongst them (britannica.com, 2023). The term was first coined in the year 2001 by the US National Science Foundation (nsf.gov, 2023) and then scientific administrators (credit goes to the famed biologist and the then assistant director of human resources and education at the NSF, Judith A. Ramaley). Meanwhile, the acronym STEM was used to refer to the curricula (that arose with the integration of knowledge from these fields) and the careers in these disciplines. In fact, thanks to her the acronym caught more than its predecessor and became a sort of household word. Subsequently, STEM-focused curricula have spread out to many corners of the world outside the US, with concurrent programs getting developed across the Asia-Pacific region (worldatlas.com, 2020) viz. in countries like France, South Korea, Australia, Taiwan, and China, and the UK in Europe.

Through STEM an inter-disciplinary approach to learning is taken where theoretical lessons are coupled with real-world instances. Teachers and pupils apply the disciplines contextually, connecting the dots between the curricula and the known world (Jindal and Tej, 2016). The Next Generation Science Standards were laid to upgrade pupils' learning comprehensibility and application of sciences, thereby accentuating the importance of STEM education. This model of education usually emphasizes project-based erudition in the classroom. The tasks, activities, and projects integrate technology to stress the application of sciences and ready scholars for their forthcoming classes (Gonzalez and Kuenzi, 2012). For the uninitiated, one can visit any reputed international institution anywhere in the world to witness this first-hand.

Pedagogy of Teaching STEM

The dynamics of STEM education are knottily tied with the world's scientific and technical advancements. The NSF upholds that pupils will need a robust foundation in STEM fields to be efficacious in this information- especially in this misinformation-driven era (sic). STEM edification and the application of science and technology empower every pupil to use their inventiveness and advance their critical rationality. The advantages of STEM education are also widespread for students of varied age groups and academic pedigree, as seen in recent years, where diversity initiatives have been envisioned and consequently minorities and females have been added increasingly to STEM fields (Sosale et al., 2023).

An important trivia to add here in the context of this chapter would be that students with ADHD (attention-deficit/ hyperactivity disorder) and other comparable issues would also benefit from STEM teachings, for instance, the ability to fulfill lengthy, nuanced tasks involving computer programming or robotics assignments and works. This point counters the age-old notion that a high attention span is needed for learning STEM subjects, and this point seconds that even in an attention-span deficit world of today, STEM education is best through experiential learning and it ought to include hands-on tasks that warrant original thinking, of course, in a conducive environment (Kumari et al., 2022).

STEM Education in Japan

Here, we give a brief overview of the STEM education ecosystem in Japan. We use Japan as our baseline for making recommendations since it has a very high level of educational diversity, particularly in the STEM fields. Since we could not locate it at the college or university levels of education, we have only presented this section from a school education angle. This topic has primarily been researched and found to be more prominent at the school level of education. However, this ought to be sufficient for our STEM ecosystem example.

Today, Japan has the eleventh-highest population on the planet. It is one of the most densely populated countries and an archipelago nation. Mountainous topography makes up around three-fourths of the nation, concentrating its densely populated urban areas. Japan has made a tremendous contribution to science and technology. Since World War II, it has become the world leader in the automotive, robotics, and electronics industries. Japan is also well known globally for its work ethic, cultural elements (art, cuisine, film, and music), and pop-cultural elements (manga, anime, and video games). Yamamoto et al., (2010) showcase how and in what ways actualized and embodied culture capital exercise affects educational outcomes in the Japanese context. The Ministry of Education, Culture, Sports, Science, and Technology (MEXT) of Japan is the authority for education in the country (wern.wes.org., 2021).

In terms of performance and quality in mathematics, reading literacy, and the sciences, Japanese pupils routinely outdo the rest of their OECD counterparts. With an average student score of 520 on the Programme for International Student Assessment (PISA) tests, (nces.ed.gov., 2022) in comparison to the OECD average of 488, Japan is a top-performing OECD nation in reading literacy, sciences, and mathematics, ranking seventh in the world in 2021-22 (oecd.org., 2022). The Japanese are highly educated, and the country puts great importance on education as a means to achieve socio-economic agility and find jobs in the high-tech sector of the economy. Policies exist to promote collaboration among teachers of specialized and regular programs (Forlin et al., 2015). Adults with post-secondary education in the nation, especially the ones having degrees in the sciences and engineering, get economic and social benefits from their education (oecd.org., 2022; studyinjapan.go, 2022).

Various universities, colleges of technology, junior colleges, and special training colleges offer higher-level and tertiary education in Japan. Only junior colleges and universities are formally regarded as higher education among these four categories of schools (studyinjapan.go,2022). These colleges offer a four-year bachelor's degree program (education.jnto, 2022) as the pinnacle of higher education, and few also offer six-year professional degree programs. Public four-year institutions are divided into two categories: the 95 provincial public universities (established by prefectures and municipalities) and the 86 state universities, which include the OUJ (The Open University of Japan).

In 2010, there were 597 four-year institutions still in operation. The state's famous schools are the most alluring for pupils looking to attain top job prospects since they offer a means of openings for kids hoping to do their post-secondary studies. International recognition exists for the caliber of Japanese universities and higher education, for instance, the THES - QS World University Rankings for the year 2022 ranks The University of Tokyo at 23rd and the Kyoto University at 36th in its list of 50 Japanese universities. The QS Asia University Rankings for the year 2022 ranks the University of Tokyo at 6th, the Kyoto University at 9th, the Tokyo Institute of Technology at 16th, and the Osaka University at 18th, in its Top 40.

Given their significant contributions to robotics, transportation, and mechanical engineering, the country's science museums reflect the extent to which they have advanced their scientific prowess. The interactive general science exhibitions of the Science Museum of Tokyo aim to increase public understanding of science and technology with a focus on young people. This museum was established in April 1964 by the Japan Science Foundation. Children can play while studying science principles that are relevant to their daily lives at the five-story museum. Kitanomaru Park is not far from the museum. Ueno Park and Ueno Zoo are just next to the National Museum of Nature and Science. This initiative was started to increase people's appreciation of biology and the study of living things. It displays dinosaur skeletons as well as species that range in size from enormous to microscopic. A massive blue whale model, pandas, and the well-known Akita dog Hachiko are all taxidermized outside.

In 1972, a huge park was established by JAXA (Japan Aerospace Exploration Agency) wherein rockets, spacecraft, and myriad facilities are present to test these flamboyant technologies. There is the Meguro Parasitological Museum of Parasites, which is the only museum in the world of this kind. There are diverse macroscopical parasites (more than 45,000 parasite specimens including the longest (8.8 meters) tapeworm in animals (and humans) exhibited in impenetrable containers. Another large science museum in Japan is Miraikan (National Museum of Emerging Science and Innovation) famed for the exhibition of a humanoid robot named Asimo. Another notable feature here is the giant LED globe. The Osaka Science Museum houses a massive planetarium (3.3 meters radius). Besides, the visitors can also see thought-provoking things concerning energy and chemistry. The Nagoya City Science Museum (known for the world's largest planetarium dome) consists of three buildings with exhibits for Astronomy, Science and Technology, and Life Sciences. There is the Museum of Aeronautical Sciences with a display of aircraft located just outside the Narita airport. It houses simulators and opportunities for hands-on activity. The museum also houses an open-air observation level that lets visitors watch how planes take off and land on Narita Airport's runway. The Gas Science Centre Museum in Toyoso, Tokyo demonstrates the working of gas and its importance in our daily lives. Tokyo Water Museum specializing in Hydrology shows the importance and wonders of water.

Importance of SPMZAs in Experiential Learning

The common understanding of science, research, and scientific literacy, is usually the common goal for science parks, museums, zoos, and aquariums (SPMZAs). Schwan et al., (2014) highlight that museums, science centers, zoos, and aquariums (MCZAs) increasingly showcase science as unfinished or present contradictory material to kindle an understanding of scientific research and construal.

Research has shown the role science centers can play in boosting science literacy levels, and getting scholars enthusiastic about STEM, thereby, offering a possibility for teachers and pupils alike to promote their scientific, engineered, and technical projects. This leads to increased participation and solutions to the socioeconomic fabric of a country. Ideally, such science centers are desired to be physically present in developing countries, preferably one in each of their major cities. Such centers can also enhance learning for the teachers, along with the taught, as "learning through experience" happens for them as well at the SPMZAs, which is warranted if one goes as per the PCK (Pedagogical Content Knowledge) construct (Loughran et al., 2012).

While there are no scientific societies that get set up in the NGO/NPO format in most growing economies, efforts to establish these can be done as grassroots endeavors. However, they must be championed by eminent academicians and scientists, as other than bringing their experience and expertise on board, they can also draw inspiration for the pupils (the future of their nations).

Propositions

Before stating our propositions, given the afore two sections, one might be able to see that SPMZAs is an all-encompassing term that includes MCZAs as well. There are nations where either format of the infrastructure does not exist, and there are nations where outdated infrastructure exists (Chowdhury et al., 2020). The economics of intermittent upgrades as well as complete overhauls is quite complex, especially with the factors of political willingness, economic GDP (spending prowess), and lack of knowledge capital, at play. As a result, the dispersion of quality STEM education is challenging and the perceived quality levels of the dispersed STEM education vary greatly amongst nations. This is seconded if one sees the PISA (Programme for International Student Assessment) 2018 and WEFFI (Worldwide Educating for the Future Index) 2019 scores and the variability in those scores of the nations.

Proposition 1: Collaboration with Existing Infrastructure

In brief, we understood that SPMZAs can act as practical laboratories for STEM institutions (Japan's STEM educational ecosystem is a live example of the same). However, there is still a lack of collaboration between the two entities. Policies ought to be in place, for mandatory physical excursions from educational institutions to the nearby SPMZA, that exist. This is warranted if

STEM education desires to advance learning comprehensibility, reflection, and reasoning, which are needed as per the New Reform (Shulman, 1987). Of course, educational institutions are slow when it comes to adopting new formats of dispersing education. The reasons for it are mainly administration unwillingness, low knowledge of new resources and techniques, and low levels of adoption by pupils coupled with low levels of cascading efforts.

Proposition 2: VR and AR-based Global Dispersion

Virtual Reality and Augmented Reality have brought the world into our palms and onto our chairs. One can sit in one's own comfort space and visit any part of the world (for instance, virtual tourism) or experience a new product especially one created in waterfall mode (new residential spaces or vehicle interiors). If one could just extend one's imagination, and of course the technology, all SPMZAs could be brought to one global platform from where anybody could access from anywhere using basic VR or AR infrastructure (say, VR goggles, etc.). This cuts out the challenging economics of upgrading SPMZAs in a developing country and for unwillingness in institutional participation for excursions. This also removes barriers to quality education, for example, a student in Afghanistan can experience a SPMZA in the USA, giving the person better exposure without waiting years for the developing country's SPMZA to get upgraded to the developed country's SPMZA standards.

Proposition 3: Promoting Firms' Participation

Going with the above propositions, there are educational institutions that might still not be interested in adopting the proposition. For instance, most private institutions in India and other similar growing economies, collect substantial fees from students but do not provide quality infrastructure to them, as the demand from the students is also low, which is appended by the fact of their and their parents' low knowledge and exposure. To cut out this lacuna, there is a need for private firms' participation in the creation of platforms for VR and AR access. This would enhance the reach of the SPMZAs to the students who can experience them on their personal VR gadgets. With the NPO format of such firms, the affordability of such services and firms' profiteering from such initiatives should not be a challenge. Public-private partnerships could help create and scale such firms.

Proposition 4: Combining SPMZAs and Heutagogy

Heutagogy has been the teaching way for medical sciences education in some eminent parts of the world, including in developing countries like India. A study (Hurson et al., 2011) implicates a few traits of this method. This self-determined TL (Teaching-Learning) method combined with AR/VR-enabled SPMZAs could be the next big innovation in STEM education dispersion.

Proposed Model: STEM Education Enrichment

The proposed model as shown in Figure 1 has three tenets to it – SPMZAs, Digital Infrastructure, and STEM Institutions. Let us explain the contextual meaning of each before explaining the model.

SPMZAs – Well, one can understand it as the physical infrastructure given the scope of this paper. We already know what the acronym stands for, as we talked about it earlier in the chapter. The reason, we are calling this ellipse SPMZA and not generic physical infrastructure, is so that one does not confuse it with the physical infrastructure of educational institutions. One can view this tenet as the laboratory readily available for integration into enriching STEM education.

Digital Infrastructure – Compared to the above, we have given the next circle a more generic name, because, it should be viewed as all existing digital infrastructure in the world today. As the world continues to advance in the digital space, we did not want to define any boundaries for it in our model. Frankly, when it comes to dispersing education in today's times, the latest and futuristic digital innovations should not hold the bar.

STEM Institutions - The last circle is the most obvious, and the key entity – the STEM institutions, particularly the educational ones.

Now that we have understood each of the tenets, let us understand the model. The intersection between SPMZA and Digital Infrastructure circles would be the Enhanced Outreach ellipse. Given the digital infrastructure around us (read AR, VR, and 24*7 Internet access) the SPMZAs outreach to the public, and especially the students would easily be enhanced. The intersection between SPMZA and STEM Institutions circles would be the Enhanced Dispersion ellipse. With such practical in-essence laboratories, the dispersion level of STEM education would be greatly enhanced for the pupils. The intersection between STEM Institutions and Digital Infrastructure circles would be the Enhanced Digitalization ellipse. This means, that STEM education and its collaterals' digitalization would greatly be enhanced with increasing levels of digital infrastructure adoption by STEM institutions. As understood, none of the ellipses alone would suffice, and thus, the intersection of all three ellipses viz., Enhanced Outreach, Enhanced Dispersion, and Enhanced Digitalization, is our sweet spot, and this sweet spot is the aim of our model – STEM Education Enrichment. Of course, enhanced dispersion is subjective, and not every region in the world has faced success or failure to the same level (Skliarova et al., 2022). We have assumed an ideal level of achievement of enhanced dispersion, to construct our model. With the intersection of all three ellipses, STEM education enrichment would indeed be a success, and for the ellipses to happen, all the three tenets discussed above must collaborate in chorus. Although the phrase sounds similar, our STEM Education Enrichment in this model does not refer to the usually implied meaning in other scholarly works (Shen et al., 2019; Fogel et al., 2015).

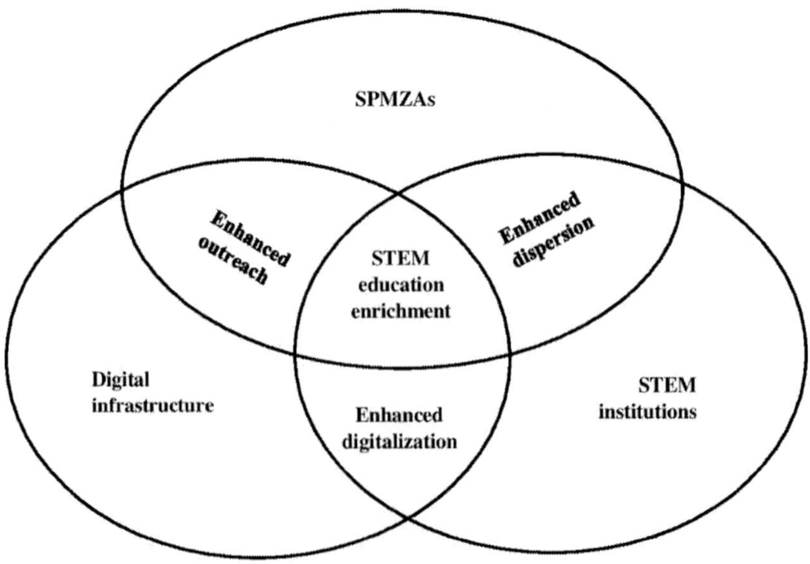

Source: Author's compilation.

Figure 1. Proposed model: STEM education enrichment.

Conclusion

We have tried to study and present a few propositions in the adoption of SPMZAs for better dispersion of STEM education in developing South Asian countries. We have also gone ahead and presented an adoptable model for enriching STEM education in these countries. We understand not all South Asian economies are currently focusing on this problem, as they have myriad diversities (Cheng, 2022), and because of this, their other challenges are of utmost priority for them – for instance, economic instability in Pakistan, Sri Lanka, military instability in Afghanistan, Pakistan, poverty-stricken Nepal, Bhutan, and worse of all in Myanmar. Today, this Northeastern neighboring country of India lags in educational standards. Factors like rote learning, lowly-trained teachers, utilization of outdated resources, and bribery are widespread at the primary as well as secondary levels of school education in Myanmar. In addition to this, pupils are compelled to follow arduous afterschool tuitions as the school erudition is considered insufficient. The pupils memorize everything, right from answers to the grammatical questions of the English language, as well as whole essays. Oftentimes the test questions are distributed in prior. Remarkably, standards 4, 8, and 10 examination responses are concealed, lest the question papers are stolen. Universities have been erected and strewn through cities to thwart students' unionization for probable unrest. We have mainly presented this keeping in mind India and Bangladesh (as these two nations have sizeable STEM education dispersion programs) and others can follow suit, once their issues are resolved.

In India, the NCSM (National Council of Science Museums), administers 26 science centers and museums pan-India. These 26 centers are distributed in North-East, North, East, South, and West zones. Amongst these 26 centers, there are 6 centers at the national level. The CRTL (Central Research & Training Laboratory) in Kolkata is the council's apex for professional training, research, and development. All these national-level centers (except the Science City in Kolkata) have corresponding regional and sub-regional (district) level centers in their ambit. These are called SUs (Satellite Units) (see Appendix A for complete details on NCSM centers and their SUs). The council has also developed museums and science centers for various provinces in the country. NCSM is also credited for creating multiple establishments for various state organizations in the likes of ONGC, BEL, and ICAR, along with collaborating worldwide for the development of museums, galleries, and science centers (e.g., Rajiv Gandhi Science Centre in Mauritius, India Gallery on Buddhism at the International Buddhist Museum in Sri Lanka, etc.). At present, NCSM has the largest network of museums and science centers in the world. India has a very full-bodied SPMZA infrastructure in the world (indianculture.gov.in). India and Bangladesh do have a robust ecosystem wherein this model could be implemented. The rest of the South Asian nations can follow suit once their other issues are resolved and stability ensues.

Disclaimer

The content in this chapter in its entirety is accountable solely to the authors and no firm or any other entity shall be held responsible. The views expressed are the authors' own views and are not on behalf of any individual or organization or any other entity in the known world.

Appendix

Appendix A: Zone-wise list of NCSM centers and their corresponding SUs

Zone	Centres	Satellite Units
East	Science City, Kolkata	
	Birla Industrial & Technological Museum (BITM), Kolkata	ShriKrishna Science Centre, Patna
		North Bengal Science Centre, Siliguri
		District Science Centre, Bardhaman
		District Science Centre, Purulia
		Digha Science Centre, Digha
	CRTL (Central Research and Training Laboratory), Kolkata	Regional Science Centre, Bhubaneswar
		Dhenkanal Science Centre, Dhenkanal and Kapilas Science Park
North	National Science Centre, Delhi	Regional Science City, Lucknow
		Kurukshetra Panorama and Science centre, Kurukshetra
		Palampur Science Centre, Palampur
West	Nehru Science Centre, Mumbai	Raman Science Centre & Planetarium, Nagpur
		Regional Science Centre, Bhopal
		Goa Science Centre, Panjim
		District Science Centre, Dharampur

Zone	Centres	Satellite Units
South	Visvesvaraya Industrial & Technological Museum, Bangalore	
	Regional Science Centre & Planetarium, Calicut	Regional Science Centre, Tirupati
		District Science Centre, Gulbarga
		District Science Centre, Tirunelveli
North-East		Regional Science Centre, Guwahati

References

Cheng, M. M. H. (2022). An Overview of STEM Education in Asia. *Concepts and Practices of STEM Education in Asia*, 1-15.

Chowdhury, S. A., Arefin, A. S., & Ahmed, F. (2020). Factors behind the implementation of STEM education in Bangladesh. In *Journal of Physics: Conference Series*, 1563 (1), 012064. IOP Publishing.

English, L. D. (2016). STEM education K-12: Perspectives on integration. *International Journal of STEM education*, *3*, 1-8.

Fogel, A., Brownlow, J., & Morgan, S. (2015). Activities in polymer optical physics for STEM education enrichment in the K-12 environment. In *abstracts of papers of the american chemical society*, washington, 249, 1155.

Forlin, C., Kawai, N., & Higuchi, S. (2015). Educational reform in Japan towards inclusion: Are we training teachers for success?. *International Journal of Inclusive Education*, *19*(3), 314-331.

Gonzalez, H. B., & Kuenzi, J. J. (2012). Science, technology, engineering, and mathematics (STEM) education: A primer. Washington, DC: *Congressional Research Service, Library of Congress.*

Herschbach, D. R. (2011). The STEM initiative: Constraints and challenges. *Journal of stem teacher education*, *48*(1), 96-122.

Hurson, A. R., Sedigh, S., Miller, L., & Shirazi, B. (2011). Enriching STEM education through personalization and teaching collaboration. *International Conference on Pervasive Computing and Communications Workshops (PERCOM Workshops)*, 543-549. IEEE.

Japanese educational system. (2022). Retrieved from https://education.jnto.go.jp/en/school-in-japan/japanese-education-system.

Jindal, P., & Tej, F. (2016). Being an educator in private vs. public school system: A comparative study on job contentment. *IRA International Journal of Education and Multidisciplinary Studies*, *3*(3), 309-319.

Judith Hallinen. (2023). *STEM education Curriculum*, Retrieved from https://www.britannica.com/topic/STEM-education on Aug 4, 2023.

Kate Boland. (2020). *10 Countries with the Best Education Systems*, Retrieved from https://www.worldatlas.com/articles/10-countries-with-the-best-education-systems.html.

Kelley, T. R., & Knowles, J. G. (2016). A conceptual framework for integrated STEM education. *International Journal of STEM education*, *3*, 1-11.

Kumari, S., Jindal, P., & Chand, P. K. (2022). Social Media Networking: As Facilitator of Acculturation. *ECS Transactions*, *107*(1), 9381.

Loughran, J., Berry, A., & Mulhall, P. (2012). *Understanding and developing scienceteachers' pedagogical content knowledge*, 12, Springer Science & Business Media.

National Center for Education Statistics (NCES) Handbook of Survey Methods. (2020). *Program for International Student Assessment (PISA),* Retrieved from https://nces.ed.gov/statprog/handbook/pdf/pisa.pdf on August 2020.

National Council of Science Museums, Indian Cultural Retrieved from https://indianculture.gov.in/MoCorganization/national-council-science-museums.

National Council of Science Museums. (2023). Retrieved from https://en.wikipedia.org/wiki/National_Council_of_Science_Museums on May 31, 2023.

Planning Study in Japan. (2022). Retrieved from https://www.studyinjapan.go.jp/en/planning/educational-system.

Schwan, S., Grajal, A., & Lewalter, D. (2014). Understanding and engagement in places of science experience: Science museums, science centers, zoos, and aquariums. *Educational Psychologist*, *49*(2), 70-85.

Science, technology, engineering, and mathematics. (2021). Retrieved from https://en.wikipedia.org/wiki/Science,_technology,_engineering,_and_mathematics on August 16, 2023.

Shen, F., Roccosalvo, J., Zhang, J., Yi, Y., Ji, Y., Guo, K., & Han, Y. (2019). STEM education enrichment in NYC. *16th International Conference on Information Technology-New Generations (ITNG 2019),* 277-282. Springer International Publishing.

Shulman, L. (1987). Knowledge and teaching: Foundations of the new reform. *Harvard educational review*, *57*(1), 1-23.

Skliarova, I., Meireles, I., Martins, N., Tchemisova, T., & Cação, I. (2022). Enriching Traditional Higher STEM Education with Online Teaching and Learning Practices: Students' Perspective. *Education Sciences*, *12*(11), 806.

Sophia Chawala. (2021). *Education in Japan,* Retrieved from https://wenr.wes.org/2021/02/education-in-japan on February 18, 2021.

Sosale, S., Harrison, G. M., Tognatta, N., & Nakata, S. (2023). *Engendering Access to STEM Education and Careers in South Asia*. World Bank Publications.

The National Science Foundation. *(2023). Improving Undergraduate STEM Education: Directorate for STEM Education,* Retrieved From https://www.nsf.gov/pubs/2023/nsf23510/nsf23510.htm on January 18, 2023.

Yamamoto, Y., & Brinton, M. C. (2010). Cultural capital in East Asian educational systems: The case of Japan. *Sociology of Education*, *83*(1), 67-83.

Chapter 10

Measuring the Effectiveness of Deep Learning in STEM Education: A Comparative Analysis of Student Outcomes and Engagement

Lalita Kumari[1,*], PhD
Chetna Priti[2], PhD
and Nikhil Sharma[1]
[1]Amity School of Engineering & Technology, Amity University, Patna, India
[2]Amity School of Business, Amity University, Patna, India

Abstract

In recent years, there has been an increasing interest in using deep learning techniques to enhance STEM education. In this study, we explore the effectiveness of deep learning in STEM education by conducting a comparative analysis of student outcomes and engagement. This study aims to measure the effectiveness of deep learning in STEM (Science, Technology, Engineering, and Mathematics) education by conducting a comparative analysis of student outcomes and engagement. Deep learning, a subset of machine learning, has shown great potential in various fields, and its application in education has gained attention in recent years. The study explores how deep learning approaches impact student learning outcomes and engagement in STEM subjects. With the help of Deep Learning student outcomes are measured through various assessment tools, including exams, projects, and performance evaluations. Engagement is assessed through surveys and observations, capturing student motivation, participation, and interest in the learning

* Corresponding Author's Email: kumaril2003@yahoo.co.in.

In: STEM
Editors: Priya Jindal, Amandeep Singh and Rashmi Aggarwal
ISBN: 979-8-89113-305-1

process. The deep learning group exhibited increased engagement, with higher levels of motivation, active participation, and interest in STEM subjects.

Keywords: STEM education, deep learning, machine learning, performance evaluation

Introduction

STEM education and deep learning are two rapidly growing areas in education that are transforming the way students learn and engage with the world around them. STEM education focuses on the integration of science, technology, engineering, and mathematics in the classroom, while deep learning is a subfield of artificial intelligence that involves the use of advanced neural networks to process and analyze complex data.

The integration of STEM education and deep learning has the potential to revolutionize the way we teach and learn. By incorporating deep learning algorithms and techniques into STEM education, we can provide students with a more personalized and engaging learning experience. Deep learning models can analyze large datasets and identify patterns and insights that may not be immediately apparent to human teachers. This can help teachers tailor their instruction to meet the individual needs and abilities of their students, providing them with more targeted and effective support.

Background and Rationale

In recent years, there has been a growing recognition of the importance of STEM (Science, Technology, Engineering, and Mathematics) education in preparing students for the challenges of the modern world. STEM education not only equips students with technical knowledge but also fosters critical thinking, problem-solving skills, and creativity. As educators strive to enhance student's learning experiences, project-based learning has emerged as a pedagogical approach that promotes active engagement and a deeper understanding of STEM concepts. Project-based learning immerses students in real-world, hands-on projects that require them to apply their knowledge and skills to solve complex problems. By engaging students in practical, meaningful tasks, project-based learning enhances their ability to think critically, collaborate, and innovate. Integrating educational robotics into project-based learning offers a unique opportunity to further enrich STEM education by providing a tangible and interactive platform for students to explore and experiment with concepts in a multidisciplinary manner. Educational robotics involves the use of robots or robotic kits specifically designed for educational purposes. These tools allow students to apply STEM knowledge in a practical context, enabling them to design, build, and program robots to solve various challenges. By engaging with educational robotics, students develop a deep understanding of engineering principles, programming logic, and computational thinking. Moreover, robotics projects often require students to work in teams, fostering collaboration, communication, and problem-solving skills.

Related Work

The studies have explored the use of deep learning in STEM education and its impact on student outcomes and engagement. Alamri and Alharbi (2021) conducted a study on using deep learning algorithms to improve student performance in physics courses. The results showed that students who received instruction with deep learning approaches achieved higher scores in conceptual understanding and problem-solving compared to traditional instruction.

Hwang et al., (2021) investigated the effects of incorporating deep learning techniques in computer science education. The study found that students exposed to deep learning methods demonstrated improved programming skills, higher creativity levels, and enhanced problem-solving abilities. In a study by Chen and Stroup (1993), deep learning approaches were implemented in mathematics education. The findings revealed that students who engaged in deep learning activities exhibited increased mathematical reasoning, conceptual understanding, and mathematical communication skills. Zheng et al., (2020) examined the impact of deep learning on student engagement in engineering education. The study found that students who experienced deep learning activities showed higher levels of motivation, active participation, and collaboration compared to traditional instruction methods. Orji and Vassileva (2020) explored the relationship between deep learning and student engagement in STEM subjects. The research indicated that deep learning approaches fostered a positive learning environment, resulting in increased student interest, curiosity, and intrinsic motivation. Deep

Learning Technologies and Tools in STEM Education (Raymundo et al., 2020) investigated the integration of deep learning technologies, such as neural networks and machine learning algorithms, in science education. The study highlighted the potential of these technologies to enhance data analysis, pattern recognition, and scientific inquiry skills among students. Mageira et al., (2022) explored the use of deep learning tools, such as artificial intelligence chatbots, in STEM education. The research demonstrated that interactive and adaptive learning platforms based on deep learning algorithms in the form of CNN and RNN as shown in Table 1, could improve student engagement, provide personalized learning experiences, and offer timely feedback. Table 1 also highlighted the significance of selecting suitable neural networks: CNNs for image analysis and RNNs for modeling time series data, aiding fields such as medical imaging and scientific trend analysis. By aligning neural network architecture with the nature of data, STEM students and researchers can enhance their analytical capabilities, improve data-driven decision-making, and advance research.

Table 1. Differentiating CNN and RNN in the context of STEM learning

Basis of Difference	CNN (Convolutional Neural Network)	RNN (Recurrent Neural Network)
Architecture	Designed for grid-like data (e.g., images)	Designed for sequential data (e.g., time series, text)
Layers	Convolutional layers, pooling layers, fully connected layers	Recurrent connections, hidden state
Data Processing	Extracts local features and spatial patterns	Captures temporal dependencies and context
Input Data	Grid-like data (e.g., images, spectrograms)	Sequential data (e.g., time series, text)
Learning Capabilities	Learns spatial features, detects patterns and objects	Captures temporal dependencies, models sequences
Applications	Image analysis, object recognition	Time series analysis, natural language processing

Source: Author's compilation.

Deep Learning in STEM Education

STEM education and deep learning can help students develop critical thinking, problem-solving, and analytical skills that are essential for success in today's rapidly changing world. Through STEM education, students can learn how to apply scientific principles to real-world problems, develop computational thinking skills, and gain hands-on experience with the latest technologies and tools. Deep learning, on the other hand, can help students develop the ability to analyze and interpret complex data, make predictions, and create models that can help solve complex problems in fields such as healthcare, finance, and engineering.

The integration of STEM education and deep learning has the potential to transform the way we teach and learn and help students develop the skills and knowledge they need to succeed in the 21st century. By providing students with a more personalized and engaging learning experience, we can help them develop the critical thinking, problem-solving, and analytical skills they need to thrive in a rapidly changing world.

Theoretical Framework

The theoretical framework for the integration of STEM education and deep learning is based on the principles of constructivism and the cognitive load theory. Constructivism emphasizes the importance of active learning and the development of critical thinking and problem-solving skills. The cognitive load theory posits that the human brain has a limited capacity for processing information and that learning is most effective when the cognitive load is optimized.

In the context of STEM education and deep learning, these theories suggest that students learn best when they are actively engaged in the learning process and when instruction is tailored to their individual needs and abilities. Deep learning algorithms can help optimize the cognitive load by analyzing large amounts of data and providing insights and recommendations that can help teachers tailor their instruction to the needs of individual students. This can help create a more personalized and engaging learning experience that promotes deeper understanding and long-term retention of information.

Conceptual Framework

The conceptual framework for the integration of STEM education and deep learning is based on the principles of active learning, personalized instruction, and data-driven decision-making. STEM education provides students with hands-on, experiential learning opportunities that promote critical thinking and problem-solving skills. Deep learning algorithms can help optimize the learning experience by providing personalized recommendations and insights based on the analysis of large amounts of data.

The conceptual framework for the integration of STEM education and deep learning also emphasizes the importance of data-driven decision-making. Teachers and administrators can use data analytics to identify areas of strength and weakness in their STEM programs and make informed decisions about instructional strategies and resource allocation. Deep learning algorithms can help identify patterns and trends in student performance data, which can be used to inform curriculum development, teacher training, and student support services (Jindal et al., 2016). The conceptual framework for the integration of STEM education and deep learning emphasizes the importance of providing students with a personalized and engaging learning experience, while also leveraging the power of data analytics to inform decision-making and improve program outcomes.

The use of deep learning models like CNNs in STEM education can help students develop skills in artificial intelligence and machine learning, which are increasingly in demand in various industries. By incorporating these technologies into STEM education, we can better prepare students for careers in fields such as computer science, engineering, and data science.

Several methodologies can be used to incorporate STEM into education. Here are a few examples:

- *Project-based learning:* This methodology involves students working collaboratively on a project that requires the use of STEM skills and knowledge. The project can be centered around a real-world problem or challenge and can involve multiple subjects. This approach encourages students to think critically, apply their knowledge, and develop problem-solving skills.
- *Inquiry-based learning:* This methodology involves students asking questions, exploring ideas, and conducting experiments to develop their own understanding of a concept. Inquiry-based learning encourages students to be active learners, to develop their own questions, and to seek out their own answers. This approach can be used to teach a range of STEM concepts, from scientific inquiry to engineering design.
- *Blended learning:* This methodology combines traditional classroom instruction with online learning resources, such as videos, simulations, and interactive activities. Blended learning can be used

to provide students with more personalized instruction and to enhance their understanding of STEM concepts. It can also provide teachers with data on student progress and allow them to adapt their instruction to meet the needs of individual students (Kumari et al., 2022).

- *Design thinking:* This methodology involves students using a structured approach to solve problems, focusing on user needs, rapid prototyping, and iterative testing. Design thinking can be used to teach a range of STEM concepts, from product design to software development. It encourages students to think creatively, work collaboratively, and develop a deep understanding of the problem they are trying to solve.

When incorporating STEM into education, it's important to choose a methodology that aligns with your teaching goals and your students' needs. It's also important to provide students with hands-on, experiential learning opportunities that allow them to develop their STEM skills and knowledge in a meaningful way.

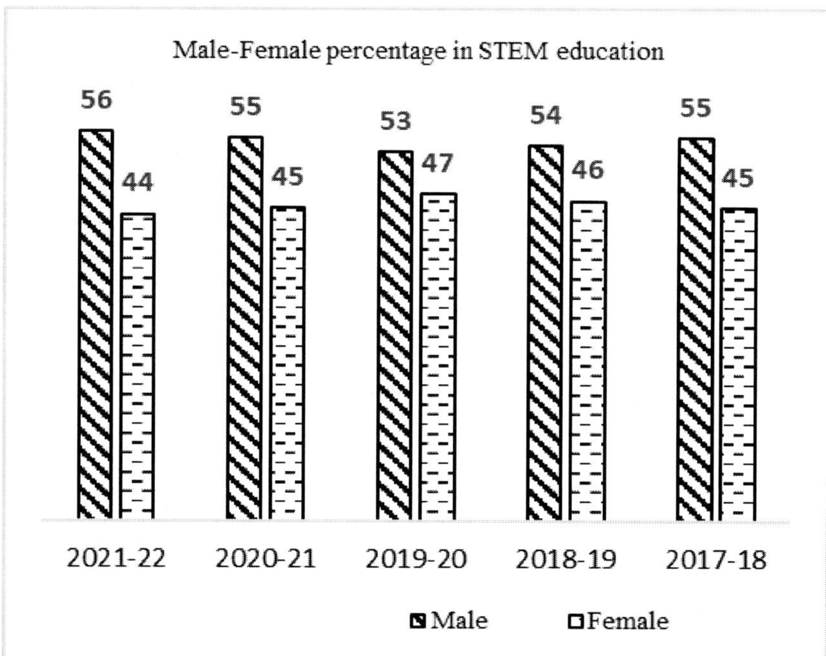

Source: AISHE and World Bank survey report (2021-2022).

Figure 1. Scholars in STEM at various labels (UG, PG, M.Phil, and Ph.D).

Figure 1 depicts STEM scholars at different education levels. More girls and women are entering STEM fields at the undergraduate and graduate levels, challenging the traditional gender norms and contributing their skills and perspectives to these disciplines.

Deep learning can be used in STEM education to enhance student learning and understanding of complex concepts. Deep learning is a subset of machine learning that uses neural networks to analyze data and make predictions or decisions. In education, deep learning can be used to develop intelligent tutoring systems, analyze student performance data, and personalize learning experiences for individual students.

There are numerous benefits of incorporating STEM (science, technology, engineering, and mathematics) education into the curriculum. Some of the key benefits are:

Developing critical thinking and problem-solving skills: STEM education encourages students to think creatively, approach problems systematically, and develop logical solutions. Fostering innovation: STEM education exposes students to cutting-edge technologies and scientific breakthroughs, inspiring them to come up with new ideas and innovations.

Preparing Students for Future Careers

The demand for STEM professionals is growing rapidly, and STEM education prepares students to succeed in a wide range of high-paying and rewarding careers.

Enhancing collaboration and teamwork: STEM education often involves collaborative projects and group work, helping students to develop important interpersonal skills and work effectively in teams.

Improving academic performance: Studies have shown that students who are exposed to STEM education tend to perform better in math and science subjects, as well as other academic areas.

Addressing real-world challenges: STEM education helps students understand and address real-world challenges, such as climate change, healthcare, and energy sustainability.

Here are a few examples of how deep learning can be used in STEM education

Intelligent tutoring systems: These are computer programs that use deep learning to analyze student performance data and provide personalized feedback and support. For example, an intelligent tutoring system might use deep learning to analyze a student's performance on math problems and provide customized hints or explanations based on the student's strengths and weaknesses.

Natural language processing: This is a branch of deep learning that focuses on understanding and generating human language. Natural language processing can be used to develop chatbots or virtual assistants that can help students with STEM-related questions or problems.

Computer vision: This is another branch of deep learning that focuses on analyzing visual data. Computer vision can be used to develop augmented reality or virtual reality applications that allow students to explore STEM concepts in a more immersive and interactive way. While deep learning has great potential for enhancing STEM education, it's important to note that it's not a silver bullet. Deep learning models require large amounts of data to train effectively, and there are concerns about the potential biases and ethical implications of using these models in education. Therefore, it's important to use deep learning thoughtfully and responsibly and to balance its use with other teaching methodologies that provide hands-on, experiential learning opportunities for students. The choice of deep learning techniques used in the study will depend on the specific research question being investigated, the nature of the STEM education curriculum, and the available resources. Here are some examples of deep learning techniques that could be used in a study measuring the effectiveness of deep learning in STEM education: Convolutional neural networks (CNNs): CNNs are often used in image and video recognition tasks and could be used in STEM education to help students visualize complex concepts, such as molecular structures or physical processes.

Recurrent neural networks (RNNs): RNNs are well-suited for processing sequential data, such as text or audio, and could be used in STEM education to help students understand complex concepts through natural language processing applications or interactive simulations. Deep reinforcement learning (DRL): DRL is a technique that uses trial and error to teach machines how to make decisions based on feedback from their environment. DRL could be used in STEM education to develop intelligent tutoring systems that adapt to student's learning needs and provide personalized feedback. Generative adversarial networks (GANs): GANs are used to generate new data that resembles an existing dataset. They could be used in STEM education to create interactive simulations or virtual environments that help students explore and experiment with scientific concepts.

Transfer learning: Transfer learning involves using pre-trained models to perform new tasks. In STEM education, transfer learning could be used to develop models that can recognize specific types of STEM-related images or audio, or to generate text responses based on specific prompts or questions.

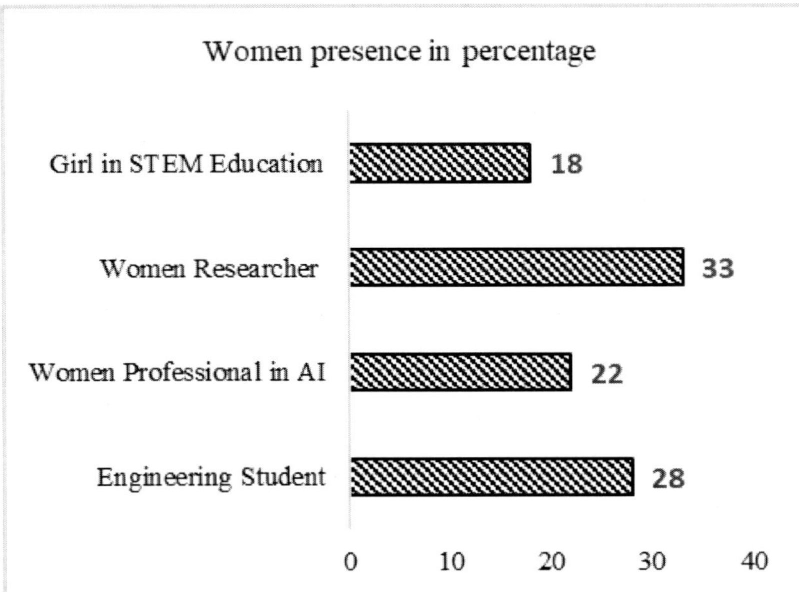

Source: AISHE and World Bank survey report (2021-2022).

Figure 2. Gender disparities in STEM education.

These are just a few examples of the deep learning techniques that could be used in a study measuring the effectiveness of deep learning in STEM education. The specific techniques used will depend on the research question and available resources.

Figure 2 describes the ratio of women and men in STEM in which women are seen as underrepresented in STEM fields and have limited participation and advancement in these disciplines. However, efforts are being made to address this gender disparity and promote greater inclusivity.

Conclusion

The study could conclude that incorporating deep learning techniques in STEM education can be an effective way to improve student outcomes and engagement. By analyzing student performance data and survey responses, the study could show that students who were exposed to deep learning approaches demonstrated higher levels of understanding and engagement with STEM concepts compared to those who were not. Additionally, the study could identify specific deep learning techniques, such as CNNs or RNNs, that were particularly effective in enhancing student learning and engagement. Furthermore, the study could highlight the importance of ongoing research and development in this area, to continually improve and refine the use of deep learning in STEM education. By doing so, we can better equip future generations with the knowledge, skills, and tools needed to tackle some of the most pressing challenges facing our world today.

References

Alamri, R., & Alharbi, B. (2021). Explainable student performance prediction models: a systematic review. *IEEE Access*, *9*, 33132-33143.

All India Survey on Higher Education, (2022). Retrieved from https://aishe.gov.in/aishe/viewDocument.action;jsessionid=58AB570C1AFE7BCF162E38A760E007F6?documentId=352.

Chen, D., & Stroup, W. (1993). General system theory: Toward a conceptual framework for science and technology education for all. *Journal of Science Education and Technology*, *2*, 447-459.

Hwang, G. J., Xie, H., Wah, B. W., & Gašević, D. (2020). Vision, challenges, roles and research issues of Artificial Intelligence in Education. *Computers and Education: Artificial Intelligence*, *1*, 100001.

Jindal, P., & Tej, F. (2016). Being an educator in private vs. public school system: A comparative study on job contentment. *IRA International Journal of Education and Multidisciplinary Studies*, *3*(3), 309-319.

Kumari, S., Jindal, P., & Chand, P. K. (2022). Social Media Networking: As Facilitator of Acculturation. *ECS Transactions*, *107*(1), 9381.

Mageira, K., Pittou, D., Papasalouros, A., Kotis, K., Zangogianni, P., & Daradoumis, A. (2022). Educational AI chatbots for content and language integrated learning. *Applied Sciences*, *12*(7), 3239.

Orji, F., & Vassileva, J. (2020). Using machine learning to explore the relation between student engagement and student performance. *24th International Conference Information Visualisation (IV)*, 480-485. IEEE.

Raymundo, C., Chavez, H., Chavez-Arias, B., Contreras-Rosas, S., & ALVAREZ, J. (2020). Artificial Neural Network Model to Predict Student Performance Using Nonpersonal Information. In *Frontiers in Education*, 8, 53. Frontiers.

World Bank Survey Report, (2022). Retrieved from https://databank.worldbank.org/US-STEM-(ISCED-and-Tertiary)/id/cd77ac48.

Zheng, R., Jiang, F., & Shen, R. (2020). Intelligent student behavior analysis system for real classrooms. *International Conference on Acoustics, Speech and Signal Processing (ICASSP)*, 9244-9248. IEEE.

Chapter 11

Employability Challenges and Opportunities under STEM Learning with a Special Focus on Apprenticeship and Internship

Praveen Nayak[1,*]
and Preeti Patil[2]
[1]Vivekanand Institute of Management, Kalaburagi, Karnataka, India
[2]Visvesvaraya Technological University, Centre for Post Graduate Studies and Research, Mysuru Campus, Karnataka, India

Abstract

Employability is a big issue with most HR Professionals and Managers agreeing that most fresh graduates, especially from traditional courses like B.Com, B.Sc., BBM, BCA, BSW, and BA, are not competent to begin work even after the induction program. The skills and knowledge they possess are mostly not industry-relevant and very few candidates have the attitude necessary to pick up necessary skills and knowledge efficiently. STEM Learning is almost unavoidable for future work. In India the new National Education Policy provides an opportunity for the University, while planning the pedagogy of all their courses, to take inputs from HR Managers, Industry Representatives, Chamber of Commerce, and MSME Department Personnel. Similarly, the UN, ILO, EU, and various Government bodies are also banking on STEM learning to address both economic growth and employability issues, which if left unresolved may lead to social problems shortly. Globally STEM is seen as the most reliable and sustainable solution to overcome the challenges brought forth by emerging technologies like AI, IoT, 3D Printing,

[*] Corresponding Author's Email: praveen16176@gmail.com.

In: STEM
Editors: Priya Jindal, Amandeep Singh and Rashmi Aggarwal
ISBN: 979-8-89113-305-1

Machine Learning, Robotics, and other beneficial but disruptive technologies (International Labour Organization, 2021; Ministry of Education, 2021). Most employed persons use only part of their STEM learning in their day-to-day work. It is also a fact that different professions and different tasks need different knowledge, skills, and experience in STEM applications. Different technology in the same industry at the same levels of work also creates different needs for STEM application for employees. This is where Internships and Apprenticeships can help individuals and specific industries design and deliver their unique STEM requirements among potential employees. While formal education syllabi are designed generically keeping in mind the overall job market requirements, the specifics KSE (Knowledge, Skill & Experience) can be delivered through Internships.

Keywords: employability, STEM learning, emerging technologies, internships and apprenticeships

Introduction

Over the years there has always been either continuous criticism of the syllabus framed by some Universities, which have failed to include industry-relevant skills in the curriculum, or praise and appreciation for the syllabus and curriculum of some institutes and Universities which have ensured their graduates are job ready with requisite skills and knowledge. Industry looks at the Knowledge, Skills, and Attitude of the job aspirants while selecting fresh graduates for their human resource. Educational Institutes and Universities and Industry must come together to assess the curriculum, from time to time, to keep the skill sets taught suitable and relevant for employability. Universities need to learn from each other as well as benchmark with the best institutions and adopt best practices.

Apprenticeships have for a long time been a system of imparting industry-specific skills. Similarly since the 1920's internships – both paid and unpaid – have also evolved as a system for transferring work-specific knowledge and skills. An apprenticeship links students with experts in their chosen field. It provides students the opportunity to learn practical skills while at the same time gaining theoretical knowledge. Through apprenticeships, the students are provided access to modern machinery, tools, and techniques that are used in the industry. This hands-on training is an essential component of the STEM education system.

Many universities and educational institutions have adopted Apprenticeships and Internships as part of their curriculum, allowing students to acquire requisite skills and get exposure to real-life experience. There also has been an increase in science, math, and technology being taught through doing – a shift from merely relying on listening, reading, writing, and reciting.

The outcome of these slow evolutionary changes was the innovative practice, proposed in 2001 as an initiative of the U.S. National Science Foundation's Education and Human Resources division, STEM. Science, Technology, Engineering, and Mathematics, i.e., STEM focused on developing essential employability skills like problem-solving, creativity, teamwork, critical analysis, communication, adopting new technology, etc.

Objectives of the Chapter

The chapter explores how STEM education has evolved in recent times and is rapidly becoming important. The benefits of STEM education, the current employment scenario, the challenges and opportunities new technologies have brought upon the employment scenario, how STEM education is being seen as a possible tool to improve employment opportunities, and how the amalgamation of internship and apprenticeship with STEM education can make STEM education more effective (Frey and Osborne, 2017). A perusal of some articles dealing with employment reveals that the situation is challenging for all nations and there is a drastic need to rapidly improve and upgrade our educational systems to prepare the manpower to be suitable for employment and entrepreneurial opportunities. Adopting STEM education is a very important step in addressing the challenges.

Review of Literature

Identifying the key employability skills: evidence from literature review Lee et al. (2020) observe the key factor for getting into a job needs additional skills apart from the subject-specific skills. Employers seek employees who have a combination of generic, technical, and job-specific skills to ensure the transfer of skills from one context to another. Employers prefer the 'right' candidate with multi-dimensional & interpersonal skills. Discussing the employability skills needed to face the demands of work in the future (Fajaryati et al., 2020) the authors observe that about 75 million youth in developing countries are unemployed. The youth unemployment rates are 2 to 4 times higher than adults. Support in the form of knowledge and skills to make them productive and able to be employed by the industry is needed. Besides technical skills, abilities in communication, collaboration, problem-solving, and critical thinking ability are sought by employers (Mahmud and Wong, 2020).

Writing about 'Graduates' employability skills' Osmani et al. (2019) tabulated a list of the most sought-after skills in Britain Communication, Teamwork, problem-solving, Technological Skills, Creativity, Interpersonal Skills, Leadership, Self-Management, Flexibility/ Adaptability, Critical Thinking, Time Management, Willingness to Learn, Planning and Organizing, Initiative, Negotiation Skills, Pre-graduation Work Experience, Working under Pressure, Self-Confidence, Personality, Independent Working Ability & Motivation, in the above order were identified in their study. The literature cited above, a small part of the immense literature available, reveals that, across the world, there is consensus that what the students learn and what the industry expects do not match well.

In today's fast-paced world, STEM (Science, Technology, Engineering, and Mathematics) education is becoming increasingly important. STEM education is more than just an acronym; it is a way of thinking that fosters creativity, innovation, and problem-solving skills. Furthermore, it prepares students for the jobs of the future by enabling them to be versed in the technical skills that will be required to power the global economy. The combination of a research-based curriculum with an apprenticeship program serves as the cornerstone of the STEM education system (Bell and Linn, 2000).

The advancements in technology have revolutionized the way we live and work. Artificial Intelligence (AI), Robotics, Internet of Things (IoT), Machine learning, and 3D printing are some of the key technologies that are transforming various industries. While these technologies have the potential to create new job opportunities, they can also lead to job losses in certain sectors.

According to reports by Accenture (Accenture, 2018), PWC CEO Report (PWC, 2017), and Deloitte Report on Industry 4.0 (Deloitte, 2018), there are many positives of these technologies, which are summarised below:

Positive Impact

1. Creation of New Job Roles

These technologies are expected to create jobs in the areas of data analytics, cyber security, product design, and engineering.

2. Increased Efficiency

AI and Robotics will increase efficiency and productivity, which may lead to the creation of more job opportunities in the long run.

3. Improved Working Conditions

With the help of AI, machine learning, and IoT, human workers can focus on tasks that require creativity and problem-solving skills.

4. Enhanced Decision-Making

Machine learning can analyze large amounts of data and provide insights that can help in making informed decisions.

5. Improved Safety and Efficiency

AI and IOT can be used to monitor and optimize work environments, leading to improved safety and efficiency.

Instances of New Jobs Created

1. Industries that specialize in AI, machine learning, and IOT will require specialized skills and knowledge, and new job roles will emerge. For example, the emergence of self-driving cars requires a new workforce of engineers, technicians, and data scientists.
2. With the growth of e-commerce, new job roles have emerged, such as e-commerce sales and marketing specialists, social media managers, and data analysts.

According to the report by McKinsey Global Institute (2017), there are many adverse impacts of these technologies, which are summarised below:

Negative Impact

1. Job Losses

The adoption of new technologies can lead to job losses in certain sectors, such as manufacturing and retail, where automation can replace human labour. McKinsey Global Institute, (2017) estimates 800 million job losses on account of automation shortly.

2. Skill Gap

The implementation of new technologies requires specialized skills leading to a mismatch between job requirements and the available workforce.

3. Economic Inequality

The benefits of technology adoption may not be evenly distributed, leading to economic inequality between the properly skilled and those who are not.

4. Social Impact

The increased use of AI and Robotics may lead to the loss of traditional jobs and the disruption of communities.

It is important to address the challenges these technologies pose and ensure that the benefits of technology adoption are fairly distributed. Governments, businesses, and educational institutions need to work together to ensure that no one is left far behind.

Further in this chapter, we will also examine how apprenticeships and internships contribute to the promotion of STEM education and how this system can improve economic growth through improved employability and the creation of jobs through entrepreneurship.

Brief History of STEM

STEM, the acronym for Science, Technology, Engineering, and Mathematics, is an interdisciplinary approach to learning that combines knowledge and skills from these four fields. The concept of STEM has a long and evolving history, which spans several centuries. The roots of STEM can be traced back to the Scientific Revolution of the 16th and 17th centuries. Scientific inquiry and experimentation led to many of the discoveries and advancements that form the basis of modern science and technology.

The 20th century saw a significant increase in the importance of science and technology in daily life. The race to space between the USA and the then USSR also contributed to rapid strides in technology. The National Science Foundation (NSF) was established in the US in 1950, which initiated the funding of science and engineering research and education (NSF, 2021).

The term "STEM" was first coined by the National Science Foundation (NSF) in the United States in the 1990s to describe the fields of Science, Technology, Engineering, and Mathematics. The term was used to describe a cohesive whole of these four fields, recognizing the importance of an interdisciplinary approach to learning and career development. Since then, the importance of STEM education has continued to grow. Today, STEM education is recognized as a critical component of global economic competitiveness, and governments and organizations around the world are investing in STEM initiatives to prepare the next generation for careers in these fields (NSF, 2021; NSB, 2020; Chubin et al., 1990).

The growth of STEM education has been accompanied by an increasing focus on diversity and inclusion. Efforts are being made to increase the representation of underrepresented groups, including women and minorities, in STEM fields.

The history of STEM education in India, the largest country in terms of population and also the largest youth population, can be traced back to the early 20th century. The rulers of the then several kingdoms recognized the importance of science and technology in modernizing the country and establishing an industrial base. The Indian Institute of Science was established in Bengaluru in 1909. After independence in 1947, the Indian government continued to invest in science and technology education. The Indian Institutes of Technology (IITs) were established in the 1950s and 1960s to provide world-class engineering education and research facilities. From the 1980s India experienced rapid economic growth, and the importance of science and technology education increased (Suneeth, et al., 2021; Biswas, 2015).

The Indian government established several initiatives to promote STEM education, including the establishment of the Department of Science and Technology in 1985 and the Science and Engineering Research Council in 1991. The government also introduced several policies to promote research and innovation in the STEM fields.

Today, STEM education is a critical component of India's development plans, and the government has launched several initiatives to promote STEM education and research. The New Education Policy (NEP) introduced in 2020 emphasizes the importance of STEM education and proposes several measures to promote interdisciplinary education in these fields. Despite many challenges, the country has made significant progress in recent years and is poised to become one of the global leaders in STEM education and research (Ministry of Education, 2021; Government of India, 2020; Aswal, 2020).

STEM in the Backdrop of Emerging Technologies

STEM education is about developing critical thinking, analytical, and problem-solving skills. It is focused on applying scientific principles to real-world problems, and this approach helps students to be more adaptable to change. The purpose of STEM education is to equip students with the technical skills necessary to succeed in the jobs of the future.

Need for Research-Based Curriculum

To ensure that students are adequately prepared for the workforce, a STEM curriculum must be research-based. This means that it is grounded in the most up-to-date scientific thinking and techniques. The curriculum should promote independent thinking, scientific inquiry, and problem-solving skills, which will aid in the development of innovative solutions (Mullis and Martin, 2017).

Global Employment Scenario

According to the International Labour Organization (ILO, 2021), the global youth unemployment rate in 2021 was estimated to be 13.7%. This corresponds to approximately 267 million young people (aged 15-24) who are not currently employed or in education or training.

The COVID-19 pandemic's impact on the global job market will be felt for some more time. According to a report by the United Nations Development Programme (UNDP), the pandemic has led to a "lost generation" of young people who face a higher risk of long-term unemployment and economic insecurity. The report notes that youth employment is likely to be one of the most pressing challenges facing the global economy in the coming years, particularly in low- and middle-income countries.

The future employment rates will depend on a variety of factors such as economic growth, education and training opportunities, and changes in the labour market. Looking further ahead, the World Bank projects the global youth unemployment rate to remain high over the next few decades, particularly in sub-Saharan Africa and South Asia. The World Bank report notes that improving education and training opportunities for young people, as well as creating more jobs in the formal sector, will be key to addressing this challenge (UNESCO, 2021; World Bank, 2021).

Efforts to Improve Employment Scenario across the World

There are a variety of measures that global bodies and major countries are taking to address unemployment issues. Here are some examples:

A. Global Bodies

The International Labour Organization
ILO has launched a Global Initiative on Decent Jobs for Youth, which aims to promote youth employment and provide support to young people seeking decent work (ILO, 2021).

The United Nations Development Programme
UNDP has launched several initiatives aimed at promoting youth employment, including the Youth Co: Lab program for Asia Pacific in association with CITI Foundation, which supports youth-led social entrepreneurship (United Nations Development Programme, 2021).

B. India

As mentioned earlier, India's National Education Policy (NEP) includes measures aimed at improving education and training opportunities for young people. The Indian government has also launched several programs aimed at promoting employment, such as the Skill India Mission, which provides training and certification to young people in a variety of skills (National Institution for Transforming India, 2021).

C. China

The Chinese government has launched several initiatives aimed at promoting employment, such as the "Made in China 2025" plan, which aims to promote high-tech manufacturing and create new jobs in the sector. China has also launched several programs aimed at providing support to small and medium-sized enterprises (SMEs), which are a major source of employment in the country.

D. USA

The U.S. government has launched several initiatives aimed at promoting employment, including the Workforce Innovation and Opportunity Act, which provides funding for job training and placement programs (U.S. Department of Education, n.d.). The U.S. government has also launched several programs aimed at providing support to small businesses, which are a major source of employment in the country (U.S. Department of Labor, 2021).

E. European Union

The European Union (EU) has launched several initiatives aimed at promoting employment, including the European Social Fund, which provides funding for education and training programs. The EU has also launched several programs aimed at promoting entrepreneurship and supporting small businesses, which are a major source of employment in the region (European Centre for the Development of Vocational Training, 2015).

F. Africa

The African Union has launched several initiatives aimed at promoting employment, such as the Agenda 2063 plan, which aims to create new jobs in key sectors such as agriculture and infrastructure.

Many African countries have also launched programs aimed at promoting entrepreneurship and providing support to SMEs.

G. South America

The Inter-American Development Bank has launched several initiatives aimed at promoting employment, including the Youth Employment Program, which provides funding for educational and training programs (Inter-American Development Bank, 2021).

Many South American countries have also launched programs aimed at promoting entrepreneurship and supporting SMEs.

These are just a few examples of the measures that are being taken to address unemployment issues around the world. The specific measures taken will depend on the economic and social context of each country or region.

Impact of STEM Education on Employment

STEM education enhances the employability of students by providing them with the necessary technical skills, which are required by employers. In addition, STEM education helps to create more job opportunities, as STEM-related fields are growing at a rapid pace. By promoting STEM education, countries are contributing to their economic growth, as STEM-based businesses contribute significantly to the global economy.

According to a report by the U.S. Bureau of Labour Statistics, STEM jobs are projected to grow by 8% between 2019 and 2029, which is faster than the average growth rate for all occupations. In addition, STEM jobs tend to pay higher wages than non-STEM jobs. For instance, in the US the median annual wage was $86,980 in 2020, compared to a median annual wage of $41,950 for non-STEM jobs (Bureau of Labor Statistics, 2021).

However, the job outlook for those who do not have exposure to STEM education may be less favorable. As technology continues to play a larger role in the job market, workers without STEM skills may find it more difficult to find employment or may be limited to lower-paying jobs.

How STEM is being Incorporated into Pedagogy across the World

STEM (Science, Technology, Engineering, and Mathematics) education is the key to building a skilled workforce that can meet the challenges of the 21st century. The integration of STEM into the education curriculum has become a global phenomenon in the last few decades. Governments around the world have recognized the importance of STEM education in preparing students for the demands of the workforce (Bybee, 2013).

STEM education has had a significant impact on employment across major nations like India, the US, the EU region, China, South America, and Africa. STEM is being adopted all across the world.

A. In India

One of the most recent examples of this is India's National Education Policy (NEP) 2020, which has emphasized the importance of STEM education from an early age. In India, STEM education has been identified as a key driver of employment growth (National Institution for Transforming India, 2021), particularly in the country's rapidly expanding technology and IT sectors. According to data from the National Skill Development Corporation (NSDC, 2021), the Indian IT industry is expected to create over 3 million new jobs by 2025, and STEM graduates with skills in areas like cloud computing, cyber security, and data analytics will be in high demand. By prioritizing STEM education, the education system is giving young people the necessary tools to succeed in a rapidly changing economy, and ensuring they are adequately prepared for the jobs of the future.

The National Education Policy (NEP) of India envisions improving employment in India through several measures, including:

1. Introducing vocational training and skill development programs that are aligned with industry needs.
2. Encouraging entrepreneurship and innovation through education and research.
3. Promoting interdisciplinary education that prepares students for a rapidly changing job market.
4. Increasing the focus on STEM education.
5. Enhancing the quality of higher education to make it more globally competitive.

India is projected to have the world's largest working-age population by 2030, with an estimated 50 crore young people (aged 15-24 years) in the country. This presents both a challenge and an opportunity for the Indian economy. If these young people can be effectively trained and educated, they could become a major source of human capital and economic growth for the country. However, if they are unable to find employment or are stuck in low-paying jobs, it could lead to social and economic problems (National Sample Survey Office, 2019; Srinivasan, 2016).

B. In the USA

The US has long been a leader in STEM education, and the impact of this focus on employment is clear. According to the US Bureau of Labour Statistics, STEM jobs in the US are projected to grow by 8.8% between 2020 and 2030, much faster than the average growth rate for all occupations. In particular, jobs in computer science, engineering, and healthcare are expected to see significant growth. The US government has made significant investments in STEM education in recent years, with initiatives like the "100Kin10" program, which aims to train 100,000 new STEM teachers by 2021, and expanded funding for STEM research and development (National Science Foundation, 2021; National Science Board, 2020).

C. In the European Union

In the EU region, STEM education has been a key priority for policymakers for many years, in part because of concerns about the region's aging workforce and declining birth rates. According to a report from the European Commission, the EU will need up to 8 million more skilled workers in the STEM fields by 2025 to keep pace with demand. As a result, the EU has invested heavily in STEM education initiatives, including programs aimed at increasing the number of women and other underrepresented groups in STEM fields (European Commission, 2021).

D. In China

China has also recognized the importance of STEM education for its economy and workforce. According to data from the Chinese government, the country is expected to require over 5 million new workers in STEM fields by 2025. In response, the Chinese government has launched several initiatives aimed at expanding STEM education opportunities, particularly in rural areas. This includes a focus on increasing the number of STEM teachers and improving the quality of STEM education in primary and secondary schools (Ministry of Education of the People's Republic of China, 2021).

E. In South America and Africa

South America and Africa have also made significant investments in STEM education in recent years. For example, in Brazil, the country's "Science without Borders" program has provided funding for over 100,000 students to study STEM-related fields abroad. In addition, the African Union has identified STEM education as a key driver of economic growth and development and has launched a number of initiatives aimed at increasing access to STEM education for students in the region (African Development Bank Group, 2021; African Union, 2021; OECD, 2016).

Overall, the impact of STEM education on employment has been significant across all these regions. STEM graduates are in high demand across a wide range of industries, including technology, healthcare, manufacturing, and finance, among others. As a result, STEM education has become an essential pathway for ensuring that individuals are equipped with the skills and knowledge needed to succeed in today's global workforce (International Institute for Management Development, 2020; NSB, 2020).

Challenges in Incorporating STEM in Curriculum

While the impact of technology on employment is complex and multifaceted, the research shows that it can have both positive and negative impacts. The adoption of AI, Robotics, IoT, Machine Learning, and 3D printing can create new job opportunities and increase efficiency, but it can also lead to job losses and economic inequality. Governments, businesses, and educational institutions need to work together (PWC, 2017) to prepare the workforce for the future and ensure that the benefits of technology adoption are evenly distributed. Integrating STEM education into the curriculum requires a strategic approach that involves collaboration among educators, policymakers, industry leaders, and other stakeholders. Some of the major challenges to integrating STEM education into the education system are:

1. Developing a Comprehensive STEM Curriculum.
2. Lack of Trained Teachers (Shekhawat, 2020).
3. Professional Development for Teachers.
4. Limited Access to Technology and Resources (Seetala et al., 2013).
5. Perception of STEM as being tough to learn (Osborne, 2003) has to be overcome.
6. Creating Industry Partnerships (NSDC, 2021).
7. Developing new and innovative Extracurricular activities related to STEM
8. Inclusion and Diversity challenges.
9. Evaluation and Assessment.
10. The Infrastructural Barriers

The challenge overall, is to integrate STEM education into the education system, which requires a collaborative effort among all stakeholders to ensure that students are equipped with the skills and knowledge needed for success in the 21st century workforce (NSDC, 2021; Singh and Singh, 2016).

Examples of the Impact of STEM on Employment

The integration of STEM education into the curriculum has had a significant impact on employment globally. A report (U.S. Bureau of Labour Statistics, 2021) suggests that employment in STEM occupations is projected to grow 8.8 percent from 2018 to 2028, much faster than the average for all occupations.

According to the World Economic Forum, the demand for STEM skills is growing at an unprecedented rate, with the number of STEM jobs expected to increase by 10.7 percent between 2017 and 2027. This demand for STEM skills is being driven by rapid technological advancements and digitization across all industries (World Economic Forum, 2018).

One of the most effective ways to bridge the skills gap in STEM industries is through apprenticeships. Apprenticeships provide hands-on training and real-world experience to students, enabling them to develop the skills required by industry. Apprenticeships have proven to be an effective way of preparing students for the workforce, with research suggesting that 91 percent of apprentices who complete their programs are employed (Helper, 2016).

The impact of STEM curriculum and apprenticeships on employment can be seen across the globe. Let's take a closer look at the situation in different regions.

A. India

The NEP 2020 has also introduced apprenticeships as a key component of vocational education. This is where the NEP is likely to be highly effective. With stress being laid on framing employability-relevant curriculum and courses designed to give practical exposures, apart from allowing students to take up some of the credits from institutes of their choice, there is bound to be an improvement in imparting job-related skills (National Skill Development Corporation, 2021).

India has the potential to become a global leader in STEM industries, with a large pool of talented engineers and scientists. However, the lack of practical skills among graduates has been a significant challenge. The introduction of apprenticeships as part of the vocational education system can help bridge this skills gap and provide students with real-world experience (Singh et al., 2018).

B. China

China has been investing heavily in STEM education and has made significant progress in recent years. According to a report by the Organization for

Economic Cooperation and Development (OECD, 2016), China is the top performer in science among 15-year-olds in the world.

The Chinese government has also recognized the importance of apprenticeships in preparing students for the workforce. The National Vocational Education and Training Plan (2011-2020) aims to provide high-quality vocational education and training to 38 million students, with a focus on apprenticeships (Xinhua, 2020).

C. USA

The US has been a global leader in STEM industries for many years, with some of the world's leading technology companies such as Apple, Google, and Microsoft headquartered in the country. The US has also been at the forefront of apprenticeship programs, with the National Apprenticeship Act (NAA) of 1937 establishing the framework for modern apprenticeships in the country.

According to the US Department of Labour (US Department of Labor, 2021), there were 633,000 registered apprenticeships in the country in 2020, with 76,000 new apprenticeship programs registered since 2017. The US government has also invested heavily in STEM education, with initiatives such as the National Science Foundation's (NSF) (Washington Post, 2019).

Increased Commitment to Incorporate STEM by Governments and Global Bodies

Governments and international bodies are increasingly recognizing the importance of STEM education in generating employment and addressing skills gaps. A cursory glance at some statistics on the amount spent, number of institutes benefiting, number of students benefited, and number of employment created through STEM education initiatives in various countries, reveals the increasing importance being attached to STEM across the globe:

A. India

The National Education Policy 2020 envisions a focus on STEM education and the development of 21st-century skills. The Skill India Mission aims to provide training and certification in STEM-related skills to millions of Indian youth. The Indian government has also launched initiatives such as Atal

Innovation Mission and Atal Tinkering Labs to foster innovation and entrepreneurship among students (NITI Aayog, 2022).

The Indian government has allocated INR 50,000 crore (approximately USD 6.7 billion) for the National Research Foundation, which will fund research and development in STEM fields. The Skill India Mission aims to provide training to 400 million Indian youth by 2022, including STEM-related skills. The Atal Tinkering Labs initiative has established over 7,000 innovation labs in schools across India, benefiting over 3 million students. The Atal Innovation Mission (Atal Innovation Mission, 2022) has supported over 10,000 start-ups and entrepreneurs in the technology and innovation space, creating over 200,000 jobs.

B. China

The Made in China 2025 initiative aims to upgrade China's manufacturing sector through the development of advanced technologies and skilled labour. The government has invested heavily in STEM education, including the establishment of over 1,000 "smart" schools and the recruitment of over 100,000 STEM teachers (China Daily, 2021).

The Chinese government has invested over CNY 50 billion (approximately USD 7.7 billion) in the "smart" education sector, including the development of STEM-related resources and infrastructure. Over 1,000 "smart" schools have been established & recruitment of over 100,000 STEM teachers has also been supported (Xinhua, 2020).

C. USA

The federal government has allocated significant funding to STEM education, including through the Every Student Succeeds Act and the National Science Foundation. The White House has also launched initiatives such as the Computer Science for All Program to expand access to STEM education and training (NSF, 2021).

The Every Student Succeeds Act provides over USD 1 billion annually in funding for STEM education programs. The National Science Foundation has awarded over USD 1 billion in grants for STEM education and research. The Computer Science for All initiative has reached over 7 million students and trained over 100,000 teachers in computer science, intending to create over 100,000 new jobs in the tech industry (OESE, 2020; NSF, 2021).

D. European Union

The European Commission has established several funding programs to support STEM education, including Horizon 2020 and the European Social Fund (European Commission, 2021). These programs aim to promote innovation, research, and skills development in STEM fields.

The European Social Fund has provided over EUR 80 billion (approximately USD 96 billion) in funding for education and skills development, including in STEM fields. Horizon 2020, (European Commission, 2021) the EU's flagship research and innovation program, has provided over EUR 80 billion (approximately USD 96 billion) in funding for research and development in STEM fields. The Erasmus+ program (European Commission, 2021) has supported over 300,000 higher education students and staff in STEM-related exchange programs.

E. International Bodies

The Global Initiative on Decent Jobs for Youth, a joint initiative of the United Nations and other international organizations, includes a focus on promoting STEM education and training (UN, 2018). The initiative provides technical assistance and funding to support youth employment initiatives, including those related to STEM.

Overall, governments and international bodies are recognizing the importance of STEM education in addressing employment challenges and promoting economic growth. However, the extent of funding varies widely across countries and initiatives, and it is difficult to provide an accurate estimate of overall spending on STEM education (Henry, 2021).

The statistics however have to be considered with caution as measuring the direct employment impact of these initiatives is difficult, as the benefits of STEM education are seen in several areas apart from direct job creation, such as innovation and economic growth (OECD, 2016; OECD, 2022).

Role of Internship & Apprenticeship in Improving Employability

Except for those in the teaching profession and scientists/researchers, most of the employed persons use only part of their STEM learning in their day-to-day work. It is also a fact that different professions and different tasks need different knowledge, skills, and experience in STEM applications. Different technology in the same industry at the same levels of work also creates different needs for STEM application for employees. This is where Internships and Apprenticeships can help individuals and specific industries to design and deliver their unique STEM requirements among potential employees. While formal education syllabi are designed generically keeping in mind the overall job market requirements, the specifics KSE (Knowledge, Skill & Experience) can be delivered through Internships. Heads of institutions and managements of institutes must be willing to invest in continuous development of skills of their staff, only then can the faculty learn and pass on the learning, STEM or any other, to the students. Necessary infrastructure must be created by both the Government and Private institutions to enhance the student's learning abilities and experiences. Apprenticeship has to be provided, wherever industries do not have sufficient opportunities, in the institutions to allow the students to get hands-on experience.

Internships and Apprenticeships in curriculum mandatorily necessitate industry involvement in suggesting curricula for various industries, which can then be incorporated by Universities and Institutions based on local opportunities (Shekhawat, 2020). Adopting Internships and Apprenticeship helps Corporate Human Resource/Managers and NGO Employees to be involved in imparting training and conducting sessions at all higher education institutions. This will drastically increase the practical knowledge and skills of the students.

STEM linked to Apprenticeship and Internship will lead to the emergence of specialized institutions to impart specific skills which would ensure that a variety of skill sets are available for students to acquire, as not all skills can be offered in every educational institute. The curriculum and infrastructure of such institutions can be based on the employment opportunities it creates.

How STEM with Apprenticeships and Internships Can Help Individuals Improve Employability

1. Career Opportunities

Students who have exposure to STEM subjects have a wider range of career options open to them. STEM-related jobs are in high demand, and these fields are expected to grow even more in the future. By incorporating Apprenticeships or Internships with STEM in the curriculum, students are allowed to gain the necessary skills and knowledge for the most coveted professions.

2. Enhanced Critical Thinking

One of the significant benefits of STEM with Apprenticeship or Internship education is that it makes students think critically and analytically. Apprenticeship or Internship involves a lot of problem-solving, which helps students develop the analytical mindset needed to solve workplace problems in the present and be ready for the future.

3. Competitive Advantage

With technology becoming more integrated into every aspect of our lives, students with Apprenticeship or Internship STEM skills have a competitive advantage in the job market. Employers are always searching for employees who understand the latest technologies and have the skills to use them to their maximum potential.

4. Higher Earning Potential

STEM jobs, post Apprenticeship or Internship experience, can fetch higher salaries than non-STEM jobs without Apprenticeship or Internship experience.

5. Innovation and Creativity

Apprenticeship or Internship clubbed with STEM education involves hands-on learning and encourages innovation and creativity. Students learn how to think outside the box and may develop unique solutions to problems, acquiring skills that are highly valued in the workplace.

6. Lifelong Learning

STEM skills acquired through Apprenticeship or Internship are constantly evolving, and those who have a solid foundation in STEM will be better equipped to adapt to new and emerging technologies throughout their careers.

Overall, including Apprenticeship or Internship with STEM education in the curriculum can help in improving employability as it equips students with the necessary skills and knowledge to succeed in a rapidly changing job market. It also helps adopt a mindset that accepts change more proactively as they experience & understand that technology changes rapidly.

Efforts to Make STEM Effective by Incorporating Apprenticeship or Internship

Globally and countries like India need to take several steps to improve the employability of its youth, including:

1. Strengthening the Education System

The education system in the countries needs to be strengthened and aligned with industry needs. This can be achieved by improving the quality of education, revising the curriculum to include more practical training especially internships and apprenticeships along with the focus on job-oriented courses, and increasing funding for research and development.

2. Encouraging Entrepreneurship

Countries can promote entrepreneurship by creating a supportive ecosystem that provides funding, mentorship, and networking opportunities to aspiring entrepreneurs. This can help create more job opportunities and boost economic growth. Apprenticeship or Internship provides an opportunity to assess one's interest in starting own enterprise in the area of work, thus boosting entrepreneurship.

3. Providing Vocational Training

Vocational training programs can equip students with the skills needed for specific industries and improve their employability. All countries should increase the availability and quality of vocational training programs to prepare students for jobs in sectors such as construction, healthcare, manufacturing, and technology apart from preparing them to pursue entrepreneurship.

4. Promoting STEM Education

STEM education can equip students with the skills needed for jobs in high-growth sectors such as artificial intelligence, robotics, and data analytics. Developing Nations must increase their focus on STEM education by improving teacher training, providing access to quality STEM resources, and investing in STEM infrastructure. Apprenticeships or Internships of teacher-educators will help in improving the quality of STEM education.

5. Encouraging Industry Partnerships

There is a need to encourage partnerships between industry and academia to bridge the skills gap and create more job opportunities. This can be achieved through initiatives such as internships, apprenticeships, and industry-academia collaborations.

6. Improving Digital Infrastructure

Improving access to digital infrastructure such as high-speed internet and e-commerce platforms can create more opportunities for remote work and entrepreneurship, especially in rural areas. Internships from remote areas would be feasible (Singh and Singh, 2019).

7. The Laws Relating to Apprenticeship

Can be suitably modified and updated to allow for proper internships and recognition of internships towards gaining educational credit.

8. Further Research in Various Areas

Like best practices of university-–industry interface, Skill needs of various industries, Manpower needs of various industries over the coming years, growth opportunities in various industries, and technological changes expected to impact various industries are needed to continuously monitor the path Employability Skills imparting takes, is needed.

By taking these steps, developing and aspiring countries can improve the employability of their youth and create a more skilled and competitive workforce.

Conclusion

The Governments and international bodies must involve all stakeholders – Universities, Colleges, Educators, Industry Professionals, Educational Experts, Policy Framers, Student Representatives and also CSR Managers, and the CSR Implementing Agencies in planning, organizing, directing, and controlling the implementation of employability skills under STEM Education.

STEM Education is a powerful tool, how well it is used as a tool to transform the input of our personnel into desired human resources will depend on how well the STEM-based curriculum is planned, designed, and implemented.

Without proper planning and integration of Internships or Apprenticeships in the STEM Education curriculum, the entire effort may end up creating a large number of youths with skills that lack demand, which is not sustainable for mankind.

Social Transformation Empathetically Materialized must be the outcome of STEM Education.

References

Aggarwal, V., and Singh, V. K. (2019). Cause-related marketing and start-ups: moderating role of cause involvement. *Journal of Global Responsibility*, 10(1), 16-30.

Accenture. (2018). The impact of artificial intelligence: Widespread job losses, or bigger productivity and wages? Retrieved from https://www.accenture.com/us-en/insights/artificial-intelligence/ai-impact-work.

Aswal, D. K. (Ed.). (2020). *Metrology for inclusive growth of India*. Springer Nature.

African Development Bank Group. (2021). Jobs for Youth in Africa Strategy (2016-2025). https://www.afdb.org/en/topics-and-sectors/initiatives-partnerships/jobs-youth-africa-strategy-2016-2025.

African Union. (2021). Agenda 2063. https://au.int/agenda2063.

Atal Innovation Mission, NITI Aayog. (2022). *Atal Tinkering Labs*. Retrieved from https://aim.gov.in/atal-tinkering-labs.html.

Bell, P., and Linn, M. C. (2000). Scientific arguments as learning artifacts: Designing for learning from the web with KIE. *International journal of science education*, 22(8), 797-817.

Biswas, A. (2015). Science education in India: Historical perspectives and contemporary concerns. *Science & Education*, 24(5-6), 475-493.

Bybee, R. W. (2013). The case for STEM education: Challenges and opportunities, NSTA Press.

Bureau of Labor Statistics. (2021). *Occupational outlook handbook: Computer and information technology occupations*. Retrieved from https://www.bls.gov/ooh/computer-and-information-technology/home.html.

Chubin, D. E., Hackett, E. J., and Hackett, E. J. (1990). *Peerless science: Peer review and US science policy*. Suny Press.

ChinaDaily. (2021). *Made in China 2025*. https://www.chinadaily.com.cn/business/2015-05/19/content_20724676.html.

Deloitte. (2018). *Industry 4.0: At the intersection of readiness and responsibility*. Retrieved from https://www2.deloitte.com/content/dam/insights/us/articles/4469_Industry-4-0/DI_Industry-4.0.pdf.

European Centre for the Development of Vocational Training. (2015). *Skills for a digital world*. Retrieved from https://www.cedefop.europa.eu/files/3074_en.pdf.

European Commission. (2021). Erasmus+. https://ec.europa.eu/programmes/erasmus-plus/what-is-erasmus-plus_en.

European Commission. (2021). Horizon 2020. https://ec.europa.eu/programmes/horizon2020.

Frey, C. B., and Osborne, M. A. (2017). The future of employment: How susceptible are jobs to computerisation? *Technological Forecasting and Social Change*, 114, 254-280.

Fajaryati, N., Budiyono, Akhyar, M., and Wiranto. (2020). The employability skills needed to face the demands of work in the future: Systematic literature reviews. *Open Engineering*, 10(1), 595-603.

Henry, M. A., Shorter, S., Charkoudian, L. K., Heemstra, J. M., Le, B., and Corwin, L. A. (2021). Quantifying fear of failure in STEM: modifying and evaluating the Performance Failure Appraisal Inventory (PFAI) for use with STEM undergraduates. *International Journal of STEM Education*, 8, 1-28.

International Labour Organization. (2021). Global Initiative on Decent Jobs for Youth.Retrieved from https://www.ilo.org/global/topics/youth-employment/decent-jobs-youth/lang--en/index.htm.

Inter-American Development Bank. (2021). Youth Employment Program. https://www.iadb.org/en/topics/youth-employment-program.

International Institute for Management Development (IMD). (2020). Global STEM Talent Competitiveness Index 2020. https://www.imd.org/contentassets/5e5f1e5c5d5f48e5aa1c5e5fa5b5c75b/2020-stem-tci.pdf.

Kumar, S. P. (2021). Impact of online learning readiness on students satisfaction in higher educational institutions. *Journal of Engineering Education Transformations*, 34(0), 64-70.

Lee, R., Raison, N., Lau, W. Y., Aydin, A., Dasgupta, P., Ahmed, K., and Haldar, S. (2020). A systematic review of simulation-based training tools for technical and non-technical skills in ophthalmology. *Eye*, 34(10), 1737-1759.

Mahmud, M. M., and Wong, S. F. (2022, June). Stakeholder's perspectives of the twenty-first century skills. In *Frontiers in Education*, 7, 931488.

McKinsey Global Institute. (2017). Jobs lost, jobs gained: Workforce transitions in a time of automation. Retrieved from https://www.mckinsey.com/featured-insights/future-of-work/jobs-lost-jobs-gained-what-the-future-of-work-will-mean-for-jobs-skills-and-wages.

Mullis, I. V., and Martin, M. O. (2017). *TIMSS 2019 Assessment Frameworks*. International Association for the Evaluation of Educational Achievement. Herengracht 487, Amsterdam, 1017 BT, The Netherlands.

Ministry of Education of the People's Republic of China. (2021). National Plan for Medium- and Long-Term Education Reform and Development (2010-2020). Retrieved from http://en.moe.gov.cn/documents/edu_prospects/201202/t20120215_129607.html.

Ministry of Education. (2021). Science, technology, engineering, and mathematics (STEM) education in India. Retrieved from https://www.education.gov.in/en/science-technology-engineering-and-mathematics-stem-education-india.

National Science Foundation. (2021). STEM Education Data and Trends. Retrieved from https://www.nsf.gov/statistics/stem-ed.

National Science Board. (2020). Science and engineering indicators 2020. Retrieved from https://ncses.nsf.gov/pubs/nsb20201.

National Institution for Transforming India. (2021). Atmanirbhar Bharat Abhiyan. https://niti.gov.in/content/self-reliant-india-mission.

National Sample Survey Office. (2019). Key indicators of employment & unemployment in India, 201718http://www.mospi.gov.in/sites/default/files/ publication_reports/ KI_Employment_Unemployment_2017-18.pdf.

National Skill Development Corporation. (2021). Skill India Mission. https://www.nsdcindia.org/skillindiamission.

NSW Government Website – Education (n.d.). https://www.education.nsw.gov.au.

Osmani, M., Weerakkody, V., Hindi, N., and Eldabi, T. (2019). Graduates employability skills: A review of literature against market demand. *Journal of Education for Business*, 94(7), 423-432.

Office of Elementary and Secondary Education (OESE), (2020) Retrieved from https://oese.ed.gov/families/essa.

Organisation for Economic Co-operation and Development (OECD). (2016). PISA 2015 Results (Volume I) Excellence and Equity in Education. Retrieved from https://www.oecd-ilibrary.org/docserver/9789264266490-en.pdf.

Organisation for Economic Co-operation and Development (OECD) (2022) https://gpseducation.oecd.org/revieweducationpolicies.

Osborne, J., Simon, S., and Collins, S. (2003). Attitudes towards science: A review of the literature and its implications. *International journal of science education*, 25(9), 1049-1079.

PwC. (2017). CEO survey: CEOs confident of growth but anxious about threats to their business. Retrieved from https://www.pwc.com/gx/en/ceo-survey/2017/gx-ceo-survey-2017.pdf.

Suneeth, B. G., Kashyap, S., Reddy, G. M., and Kaushal, V. (2021). Resilience adaptations in tourism education for the post-COVID-19 era–a study of India. In *Tourism Destination Management in a Post-Pandemic Context*, Emerald Publishing Limited, 291-302.

Shekhawat, S. (2020). Enhancing employability skills of engineering graduates. In *Enhancing Future Skills and Entrepreneurship: 3rd Indo-German Conference on Sustainability in Engineering*, Springer International Publishing. 263-269.

Srinivasan, V. (2016). *The intelligent enterprise in the era of big data*. John Wiley & Sons.

Seetala, N., Himaya, M., Derosa, P., Ware, M., Hubbard, D., and Reddy, Y. (2013). Research Experience–Stem Undergraduate Gear-Up. In *Inted2013 Proceedings* (Pp. 2792-2801).

Suman, S., Narayan, M., and Singh, M. K. Bridging Skill Gap in India: A Roadmap for Atmanirbhar Bharat.

U.S. Department of Education. (n.d.). STEM education. https://www.ed.gov/stem.

U.S. Department of Labor. (2021). Workforce Innovation and Opportunity Act. https://www.dol.gov/general/topic/training/wioa.

UNESCO. (2021). Global Education Monitoring Report. Retrieved from https://en.unesco.org/gem-report/.

United Nations Development Programme. (2021). Youth Co: Lab. https://www.youthcolab.org.

World Bank. (2021). Science, Technology, Engineering, and Mathematics (STEM) Education in Asia: A Comparative Review. https://openknowledge.worldbank.org/handle/10986/32187.

Xinhua. (2020). China Focus: "Smart" education sector sees accelerated development in China. Retrieved from http://www.xinhuanet.com/english/2020-09/18/c_139379053.html.

Chapter 12

Moving Towards Practicality in Higher Education through STEM Education and Work-Based Learning

**Akshita Chotia*
and Sneha Thakur**
Department of English, Poornima University, Jaipur, India

Abstract

Work-based learning model in the arena of human resource development has garnered a considerable amount of attention in the realms of higher education in recent years. It serves people who are involved in decision-making and organizational behavior for policymakers, managers, and the developers of the curriculum. In the global economy of the 21st century STEM- Science, Technology, Engineering, and Mathematics have entered the race of assuming economic importance and becoming global. It has become a meta discipline by surpassing the cages of traditionalism running down the lines of innovation and designing solutions to complex contextual solutions with current technologies and tools. For the success of STEM-based education, it is important for both the parties – the teachers and the students to be ready to receivewhat it has to offer – a well-planned curriculum, mediums of apt technology, and engineering. Thespirit has to be included in the curriculum of science and mathematics to promote scientific inquiryand engineering design processes. Educators all over the community can work towards the formulation of apt pedagogical models that support the understanding of STEM instruction. The proposed study aims to understand and propagate the efficacy of

* Corresponding Author's Email: akshitasharma0023@gmail.com.

In: STEM
Editors: Priya Jindal, Amandeep Singh and Rashmi Aggarwal
ISBN: 979-8-89113-305-1

STEM in particular and work-based learning in general. It also seeks to define the necessary attributes of STEM and its further progress in terms of third-world nations like India.

Keywords: STEM, work-based learning, pedagogy, higher education, pedagogical models

Introduction

Since the advent of globalization and modernization, work-based learning has become a very important educational strategy which is to keep providing students with hands-on training experiences that can enrich opportunities in the domain of academic and technical studies.

It is a series of courses that integrate the curriculum of the school with the needs of the industry, creating a different learning paradigm. It is an attempt to deliberately merge theory with practice and acknowledge the intersection of explicit and tacit forms of knowing. The Work-based learningprograms are focused on quality assessment as per UGC norms, which is providing quality to thelearners and focusing on the fulfillment of the industry requirements. The existing programs intendto bridge the lacuna between learning and doing (Mutakinati, et al., 2018). It creates career awareness, career exploration, and planning, and helps students attain employability competencies and positive work attitudes. It provides arrangements including internships, placements, and informal learning on the job. The driving force is the need to meet the workplace. Work-based learning programs are classified in three ways – Duration of assignment (a few hours – to 2/4 years), Relation to course work (related to a specific subject), and Stipend (Paid or unpaid assignments). The work-based learning strategies include apprenticeships/internships, job shadowing, industry trips, entrepreneurial experience, cooperative education, school-based enterprise, and service learning (Esters and Retallick, 2013). The key people who are involved in this process are – the student, the business mentor, the teacher coordinator, the school administrator, and the parents. After the cooperation of all, the initiative can be successful.

There are innumerable benefits of work-based learning. Firstly, the student and the teacherget a reality check of the application of the things that are taught at the classroom level. The learners can understand the relevance of their education in comparison to the needs of the university. This promotes the ability to think critically and provide analytical reasoning and logical abilities.

The academia will also be able to identify gaps between what is taught and what should be taught, thus leading to the expansion of the curriculum, finally doing justice to the diverse needs of the learners by making them aware of their career opportunities. It will also createa pool of skilled and talented future employees (Fergusson et al., 2020). A noticeable difference will be seen in the time devoted to training and cost. This will make education relevant and valuable to the social context,and benefit the productive economy.

However, one might also consider the disadvantages of the initiative of Work-Based Learning programs. It consumes time to identify the kind of courses that can be taught during these programs. It needs a tremendous amount of planning to introduce or adjust Work-Based Learning courses in the existing curriculum, as not all courses would be in sync with the timelines of formaleducation. An effective level of coordination between all the key persons involved is necessary. Furthermore, an evaluation strategy needs to be developed for the assessment of student performance, involving summative and formative feedback. It is the need of the hour that educationincreases its visibility by showing its links to the industry (Jindal and Tej, 2016). Today, it is safe to say that all fields ofstudy can follow the work-based learning methodology. In India, science education is somethingthat has a lot of scope and with its practicality, is beneficial to society as well. Students, having good academic records take up sciences after 10 + 2. After completing their 12th with sciences, a huge chunk moves towards pursuing an engineering degree or a medical degree. The reason is veryclear- employability. Then, we have a very minor portion of these science students who take up aBachelor of Science – B.Sc degree – either because they are interested in further studies and research or are preparing for something else. The state of science education in India is not very good. The inability of academia to join hands with industry leads to a lot of good students dropping out as they do not see a point in continuing with it. STEM education is an acronym that is usedto show the amalgamation of four disciplines – Science, Technology, Engineering, and Mathematics. The basic of STEM education differs from the regular learning of science as it is more focused on the practical application of the theory that is taught in the class, thus- integrating applied approach and cohesive learning techniques.

STEM Education

'STEM' first appeared at one of the programs of the National Science Foundation (NSF) in 1993 proposing improvement in the quality of the USA's science, Mathematics Engineering, and Technology education. However, in 2001, Judith Ramaley reordered the words to emphasize the interdisciplinary emphasis as the new order suggests a meaningful connection between them. It isnow a frequently used acronym in the world of education that combines the essences of the four discrete disciplines for the development of the technology and skills required to work on the processes with creativity and the importance of need-based and practical solutions. In today's competitive world, education is an important determiner of one's economic stand and the economic growth of a particular nation. Prior research has also helped us draw parallels between the positive association between schooling and economic growth. The above defines a mechanism that has been important in identifying the theoretical ground of literature which is also impacting education. Education is instrumental in increasing productivity and outcomes with the integration of human capital and labor force. Secondly, with its continuous contribution to the storehouse of existing knowledge in society, it also helps in the enhancement of knowledge regarding new technologies, products, and the process of implementation. Third, education is becoming a facilitator of the diffusion and transmission of knowledge for a better understanding of existing technologies. The effect that education has on economic growth has been combined withmany other factors like variations in cognitive skills resulting from contributions from family, peergroups, and health and education.

The latest research has testified to the efficacy of having a creative and problem-solving attitude in the context of the innovation of new technologies which is working for the success of creating future economy and jobs. A decline in enrolment in STEM education, stemming from a lack of interest in the field combined with the rising global environmental problems has urged the policymakers to take serious steps in invoking interest in the targeted student demographic for STEM. The Indian government has remained committed to inspiring and training the youth towards the creationof a skilled workforce that shall prove to be invaluable contributions to the STEM industry. The vision shall be realized with the provision of innovative education that parallels the interests,skills, and achievements of students with the proper amalgamation of a team of dedicated teachers. Bythe usage of the methods of descriptive case study, we shall try to understand the growth of STEM education in India, via recent literature available, policy documents, and India's approach and philosophy towards education. Work-based learning model in the arena of human resource development has garnered a considerable amount of attention in the realms of higher education inrecent years. It serves the purpose of people who are involved in decision-making and organizational behavior for policymakers, managers, and the developers of the curriculum (Hutchinson, 2012).

In the global economy of the 21st century STEM- Science, Technology, Engineering, and Mathematics have entered the race of assuming economic importance and becoming global. It has become a meta discipline by surpassing the cages of traditionalism running down the lines of innovation and designing solutions to complex contextual solutions with current technologies and tools. For the success of STEM-based education, it is important for both the parties – the teachers and the studentsto be ready to receive what it has to offer – a well-planned curriculum, mediums of apt technology,and engineering. The spirit has to be included in the curriculum of science and mathematics to promote scientific inquiry and the engineering design process. The vision of STEM Literacy will be understood when all teachers are provided proper guidance so that they can guidethe students further in the process. Educators all over the community can work towards the formulation of apt pedagogical models that support the understanding of STEM instruction. The proposed study aims to understand and propagate the efficacy of STEM in particular and work-based learning in general. It also seeks to define the necessary attributes of STEM and its further progress in terms of third-world nations like India. Education is highly valued in India; it isdeeply rooted in the philosophy of the country and it is evident everywhere. The Indian system of education and policies are based on a philosophy curated in the Vedic times system '*Swadeshe pujyate raja, vidwan sarvatra pujyate*' that translates to 'a king is honored only in his own country,but one who is learned is honored throughout the world.' The 10 + 2 system is based on the recommendation of the Kothari Education system of 1964-66 where the education consists of primary(for 6-10 year old) and upper primary (11-14 years old), secondary education (14-16 years), seniorsecondary (16-18). After the advent of RTE (Right to Education), free schooling is provided and is compulsory for all children between the age bracket of 6-14 years.

Current Situation of STEM Education

India has had a history of commendable scholars like *Charaka, Susruta, Aryabhata, Bhaskaracharya, Chanakya, Patanjali, and Vatsyayana* who made Science and technology very advanced. Engineering and technology are a part of the regular curriculum of pedagogy. Three potholes have been identified in science education in India: unavailability and lack of access to basic infrastructure and scientific equipment for teaching the subject, lack of teachers who are updated with the latest information and education in the field, and lastly, the curriculum of the subject that is followed all across the country. Indian education system focuses more on providingquality education to the students. It focuses on establishing a bond between the student and cooperation. Since time immemorial, society has focused on the growth and development of the nation. As long as we are talking in terms of education, we are focusing on bringing up the schemes, which promote the quality of the students. Ministry of Skill Development is also comingup with certain schemes that focus on the holistic growth of the child in terms of providing a vocational and practicality-based education system. With the advent of modernization and globalization, we are in a world of technology where rationality has taken over technology (Kumari et al., 2022).

STEM Education focuses on bringing science education with the proper blend of other functional streams which has become very important for the growth and development of the childin general. The Indian system of teachers has the 'Guru-shishya concept' where the parties share a very metaphysical relation. The teacher is not just a teacher, he is elevated to the status ofa deity. We need an infallible army of teachers who could be the 'gurus' of the next generation of 'shishyas.' Teaching as a career is not the first choice of science and math graduates. They prefer working in the field, rather than working as teachers. Thus, there is a paucity of the number of well-qualified teachers who have command over their subjects in ratio to the number of students who want to learn.

The infrastructural quality is far from satisfactory. Teaching science needs the apt prerequisite amount of academic knowledge, a pre-service training to deal with classroom and laboratory queries. Thus, to fulfill the criteria and provide a dedicated training period for would-be teachers, the RTE Act and NCTE (National Council for Teacher Education) have made undergraduate degrees along with a two-year B.Ed. course as an entry-level qualification to ensurebetter quality teachers. Until then the system still consisting of not-so-proficient teachers from yesteryear's dedicated training programs and workshops along with strict monitoring of the pedagogies and teaching methodologies can be considered as corrective measures. STEM education is helping students to go beyond their capacity and reach heights. This has given thestudents a Universal platform which is very necessary to compete with international students. Technology is a global issue now. It is no longer just a domestic thought to talk about. STEM education is coming up with many scholarships that are helping students to explore more withoutany financial burden. It is also allowing students to showcase their talent on international platforms. The students of remote areas are also eagerly participating in the growth of this industry.

Skill Development Programs

Another comprehendible reason why students are quitting the science field as it has somehow failed to make linkages with the vocational fields. The academic accomplishments of someone who graduates in this field remain unnoticed by the industry and thus, no avenues are provided for free mobility in India. According to the vision of the newly formulated comprehensive national policy for skill development programs, twenty-five percent of the schools in the skill developmentprograms by 2022, projected themselves involved in 'Skill India.' The Rashtriya Madhyamik Shiksha Abhiyan (RMSA) revamped itself as 'Vocationalisation of Higher Secondary Education' during the twelfth plan for imparting skills to the students. Under this scheme, 0.15 million students across 16 states, encompassing three thousand schools have been trained. The scheme is in line with theongoing Australian school VET vocational education training, which has been training high school teachers for decades in different vocational skills. The relevance of the curriculum is seen in itsprovision of a bucket of paths after school including universities, apprenticeships, and employment.

STEM for School Initiatives

The government of India is working hard to evoke the temperament for science among students through the inclusion of various interesting programs and the provision of research training opportunities for the creation of a better workforce. The major science and technology programs for Indian school students are as follows.

National Children's Science Congress

National Children's Science Congress (NCSC), initiated in 1993 by the government of India is anationwide flagship program of the Department of Science and Technology (DST). DST was an apexnodal department of the Government of India in May 1971 for the organization, promotion, and coordination of science and technology activities in the country. NSCC promotes the submissions of open-ended scientific projects made by young minds working individually or in teams. Since its commencement, approximately one million children under the age group of ten toseventeen have been provided a platform to address the problems of society through science and technology. In today's competitive world, education is an important determiner of one's economic stand and the economic growth of a particular nation. Prior research has also helped us draw parallels between the positive association between schooling and economic growth. The above-defined mechanisms have been important in identifying the theoretical ground of literature which is also impacting education. Education is instrumental in increasing productivity and outcomes with the integration of human capital and labor force. Secondly, with its continuous contribution to the storehouse of existing knowledge in society, it also helps in the enhancementof knowledge regarding new technologies, products, and the process of implementation. Third, education is becoming a facilitator of the diffusion and transmission of knowledge for a better understanding of existing technologies. The effect that education has on economic growth hasbeen combined with many other factors like variations in cognitive skills resulting from contributions from family, peer groups, and health and education.

Linkages with Tertiary Institutions

The Indian Education system proudly stands as the world's largest higher education that includestechnical education with seven hundred ninety-nine universities, 39071 colleges (as per the annualreport, 2014-2015), and eleven thousand nine hundred twenty-three stand-alone institutions in India. The gap between the education that is provided in schools and its future utilization in tertiary institutions has led to a demotivation of students. As we are progressing towards betterment in the field of science and technological education, we have to put our focus more on bringing different schemes that can be availed by the students and organizations to fetch the attention of other students towards science-based education. Teachers' professional development encourages teachers to propagate and share methodologies on innovation in the domain of science.The seven NTSC organized biannually has witnessed enthusiastic participation on the part of teachers. NCERT and the Department of Education provide pre-service and in-service training for teachers in the field of science and mathematics. Ranked virtually at the bottom of the PISA study of seventy-four countries in terms of scientific and mathematical literacy, India has significant scope for cooperation from various universities abroad that are equipped with established joint research facilities. India, with its properly devised planning, can aim to collaborate with other countries, that might still be at the nesting stage or above. Such collaborations would serve as an assetfor STEM education development in India as they would open ways for collaborative research and development projects.

Expenditure on Education

The Indian national education policies (National Policies on Education, 2016) have promised around 6% of the GDP as minimum expenditure on education and only 0.88% towards research and development as revealed by the research and development statistics, 2011-12. Education is a concurrent subject that was earlier a state subject (Sharma and Yarlagadda, 2018). It needs funding from the center and stateside. Education is very important for generating human capital and for the growth of the nation. Science-based education with a blend of innovations will bring a lot of capital to the country in terms of innovations. India in the upcoming years will stand strong in terms of science-based start-ups. It willfetch a lot of people to showcase their talent and to provide our country with a list of innovations.Therefore, the major focus of the government should be on providing education-based schemes that can allow financially burdened people of remote areas. The more united we stand more we will be able to fetch the conscience of the minds of people.

Shortcomings

The Indian education system, which has so far revolved around cultures, and morals, added by the many shifts of the reigns of various rulers, lacks the focus and competency-based approach that isfollowed in the West and well suits the need for STEM education. Despite the right to free education and compulsory education in 2009, it is facing challenges as compared to other nations, because of their cultural diversity and the size of the population. Mathematics and Science have adifferent frenzy in India, owing to societal concerns. Parents push their children to take up sciences in their higher education. Reforms in the curriculum of science, inept infrastructure, below-satisfactory teachers, specializing in STEM, and lack of specialization for the identificationof early talents are important for the improvement of STEM education in India. The estimated shortage of 0.5 million teachers in elementary schools- 14% of secondary government schools donot have a minimum of six teachers. These vacancies are not sought as teachers are unwilling to workin rural areas as these areas do not provide proper facilities. This is leading to a large proportion of low achievers. These are to be resolved by better communication between the administration and the teachers. Less and rigid options in numeracy also account for increased rates of failure and studentsdropping out from early high school. The Indian education system doesn't allow students to choose the difficulty level in which they want to write their Xth board examinations. However, thenational education policy (2016) also recommends two levels: Part A at a higher level and Part B atalower level for the 10th board examination in Science and Mathematics to give students the to exercise their choice (Sharma and Yarlagadda, 2018).

Vocational Education and Training

The Indian government, keen on fulfilling the Skill India program, has accepted the damage that the lack of vocational courses has caused and also the pathways in which one may enter the STEM streams. Except for a few certificate and diploma courses after the 10th or 12th, there were no noticeable programs for vocational training in India. coordinates creates, and augments the training deliverycapacity. These ministries achieved 8.6 million youth during 2014-2015. The honorable prime minister of the nation has time and again emphasized the fact that progress in the field of science, technology, and innovation, will reflect the progress of the nation. The PM asked for the identification of young scientific brains among school-going children and effective mechanisms for the reduction of drudgery through the promotion of science and technology by 2022. The constructivist teaching model plans to spend a major portion of its research budget on STEM. Undoubtedly, the implementation of programs like NCSC, Science Express, and INSPIREhelps to increase awareness about science and its prospective career options. However, it is not devoid of shortcomings when these STEM programs are targeted at schools. Only the top one percent of the well-performing students are selected, the rest, despite having an aptitude for science,are left behind. Repetitive participation of favored students and schools in science programs like NCSC and IRIS discourage the participation of those who are genuinely interested. Also, such programs ought to have a better reach to the tertiary institutions, which is not true in this case.

India has strong chances of linkages of the best stem practices for students, teacher trainingand exchange as well as vocational education. There stands a dire need for a skilled workforce, a dedicated duo of teachers and students, proper infrastructure, challenges of population size, languagediversity, and lack of trained teachers. There are innumerable benefits of work-based learning. Firstly, the student and the teacher get a reality check of the application of the things that are taught at theclassroom level. The learners can understand the relevance of their education in comparison to the needs of the university. This promotes the ability to think critically and provides analytical reasoning and logical abilities. The academia will also be able to identify gaps between what is taught and what should be taught, thus leading to the expansion of the curriculum, finally doingJustice to the diverse needs of the learners by making them aware of their career opportunities. It will also create a pool of skilled and talented future employees. A noticeable difference will be seen in the time devoted to training and cost. This will make education relevant and valuable to the social context, and benefit the productive economy.

The redesign, revamp, and remodel of the education system is the need of the hour. This can be done by curating an effective process for selecting, promoting, renumerating, and proper transfer of teachers (Bozkurt and Tan, 2021). A healthy system facilitating the exchange of teaching tools, methodologies, and resources for training teachers including trainers would significantly improve the quality of the Indian education system. India can be an emerging knowledge economy. Sharing best practices through a healthy dialogue may help fulfill the vision between the nations. Working mutually, keeping in view, the mutual benefits and interests of the educational authorities, the universities, and the industries are important for the effective execution of their competing programsand resources for STEM education (Kim et al., 2018). The expansion of Work Based Learning shouldn't stop on STEM; it should expand its scope towards the fields of arts and commerce as well, as fields of study that identify under it. Everybody who is pursuing education should have an equal scope ofemployability and learning, beyond books. If this vision is realized, then surely, India will be an economy, where almost everyone, who has undergone the phase of education and employability exercises will be an equal and proud contributor to its glory and success.

Future Scope

STEM education has become an umbrella term for the rest of the streams. In the upcoming years,it will have a more massive impact on education and work-based learning. This science-dominated education is intended to bring a major revolution in the Indian education system. STEM-leadingeducation will bring effective changes in the generation of Employment. Education and employment together will create the desired capital for the nation. It will also fetch students in the domain of science and will promote start-up base innovations which is a basic requirement to present India on the international level. In today's competitive world, education is an important determiner of one's economic stand and the economic growth of a particular nation. Prior research has also helped us draw parallels between the positive association between schooling andeconomic growth (Lee et al., 2019). The vision of STEM Literacy will be understood when all teachers are provided proper guidance so that they can guide the students further in the process. Educators all over the community can work towards the formulation of apt pedagogical models that support the understanding of STEM instruction. The proposed study aims to understand and propagate the efficacy of STEM in particular and work-based learning in general. It also seeks todefine the necessary attributes of STEM and its further progress in terms of third-world nations like India. Education is highly valued in India; it is deeply rooted in the philosophy of the country and it is evident everywhere.

As STEM is progressing at a rapid pace, there is a dire need to create an equal workforce to nurture the idea of this science, technology, engineering, and mathematics-based education (Baran et al., 2016). It willhelp in achieving the multidimensional educational goal through standard-oriented teaching. The co-development of science and math will allow Indian students to represent India on international grounds. The Indian government has remained committed to inspiring and training the youth towards the creation of a skilled workforce that shall prove to be invaluable contributions tothe STEM industry. The vision shall be realized with the provision of innovative education that parallels the interests, skills, and achievements of students with the proper amalgamation of a team of dedicated teachers. By the usage of the methods of descriptive case study, we shall try tounderstand the growth of STEM education in India, via recent literature available, policy documents, and India's approach and philosophy towards education. The rationality of these subjects will help with the holistic development of the student. The properly designed course and the support of management are crucial for the fulfillment of the existing target in terms of futuristic scope.

Conclusion

To conclude, the present research paper focuses on the importance of a better linkage of work-based learning with productive science-based learning education. The era of globalization has focused on the growth of technology on the major front. It has developed technology-based education overthe years. Work-based learning skills can prove to be very beneficial for the holistic developmentof the students of the science stream. STEM education is helping students to go beyond their capacityand reach heights (Rifandi and Rahmi, 2019). This has given the students a Universal platform which is very necessary to compete with international students. Technology is a global issue now. It is no longer just a domesticthought to talk about. STEM education is coming up with many scholarships that are helping students to explore more without any financial burden. It is also allowing students to showcase their talent on international platforms. STEM education, if focused properly can bring positive changes in the existing education system. Science, technology, engineering, and mathematics are such strong fields if combined with proper schemes can bring the desired changes in the field of education. Teaching science needs the apt prerequisite amount of academic knowledge, a pre-service training to deal with classroom and laboratory queries. The need of the hour is the proper implementation of existing schemes and the focus on infrastructure development, which can benefit students, in the long run, to achieve their targets in the domain of STEM innovations. The properly designed course and the support of management are crucial for the fulfillment of the existing target in terms of futuristic scope. With the rapid pace of Westernized education, this science-based learning will help students to grow more beautifully. The above-mentioned future scope can bring a revolutionary change in the field of Indian education. The better representation of the students will create more human capital and will make the country self-sufficient. This self-sufficiency will allow the negative externalities to fade away and will create Aatmanirbhar human capital. The rationality of these subjects will help with the holistic development of the student. The aforementioned facts have focused on the idea that Work-based learning with the amalgamation of STEM education will bring the revolutionary desired changes.

References

Baran, E., BİLİCİ, S., Mesutoglu, C., & Ocak, C. (2016). Moving STEM beyond schools: Students' perceptions about an out-of-school STEM education program. *International Journal of Education in Mathematics Science and Technology*, *4*(1).

Bozkurt Altan, E., & Tan, S. (2021). Concepts of creativity in design based learning in STEM education. *International Journal of Technology and Design Education*, *31*(3), 503-529.

Esters, L., & Retallick, M. (2013). Effect of an experiential and work-based learning program on vocational identity, career decision self-efficacy, and career maturity. *Career and Technical Education Research*, *38*(1), 69-83.

Fergusson, L., Shallies, B., & Meijer, G. (2020). The scientific nature of work-based learning and research: An introduction to first principles. *Higher Education, Skills and Work-Based Learning*, *10*(1), 171-186.

Hutchinson, J. (2012). Career-related learning and science education: The changing landscape. *School Science Review*, *346*, 91-98.

Jindal, P., & Tej, F. (2016). Being an educator in private vs. public school system: A comparative study on job contentment. *IRA International Journal of Education and Multidisciplinary Studies*, *3*(3), 309-319.

Kim, N. J., Belland, B. R., & Walker, A. E. (2018). Effectiveness of computer-based scaffolding in the context of problem-based learning for STEM education: Bayesian meta-analysis. *Educational Psychology Review*, *30*, 397-429.

Kumari, S., Jindal, P., & Chand, P. K. (2022). Social Media Networking: As Facilitator of Acculturation. *ECS Transactions*, *107*(1), 9381.

Lee, M. H., Chai, C. S., & Hong, H. Y. (2019). STEM education in Asia Pacific: Challenges and development. *The Asia-Pacific Education Researcher*, *28*, 1-4.

Mutakinati, L., Anwari, I., & Kumano, Y. (2018). Analysis of Students Critical Thinking Skill of Middle School through STEM Education Project-Based Learning. *Jurnal Pendidikan IPA Indonesia*, *7*(1), 54-65.

Rifandi, R., & Rahmi, Y. L. (2019). STEM education to fulfil the 21st century demand: a literature review. *Journal of Physics: Conference Series*, 1317 (1), 012208. IOP Publishing.

Sharma, J., & Yarlagadda, P. K. (2018). Perspectives of 'STEM education and policies' for the development of a skilled workforce in Australia and India. *International Journal of Science Education*, *40*(16), 1999-2022.

Chapter 13

Frugal Innovation: Lessons from Successful Entrepreneurs and Trends Shaping STEM Future

Garima Bansal*, **PhD**
Department of Commerce, Hindu Girls College, Sonipat, Haryana, India

Abstract

Frugal innovation is an emerging concept that has been gaining increasing attention in recent years, particularly in emerging markets. It involves creating innovative products, services, or business models with limited resources and has been used successfully by many entrepreneurs to create profitable businesses. This chapter proposes to provide a comprehensive overview of frugal innovation, including its definition, relevance in the global economy, and key success factors. The chapter will focus on case studies of successful entrepreneurs who have utilized frugal innovation techniques, examining their strategies and challenges, and highlighting valuable insights for entrepreneurs seeking to adopt similar approaches. Furthermore, this chapter will discuss the trends shaping the future of frugal innovation, including emerging technologies, changing consumer demands, and new business models. It will also address the role of education and government policies in promoting entrepreneurship development and frugal innovation. The chapter will conclude by providing practical guidance for entrepreneurs and policymakers on how to foster frugal innovation and promote sustainable economic growth.

---

* Corresponding Author's Email: garimahgc13@gmail.com.

In: STEM
Editors: Priya Jindal, Amandeep Singh and Rashmi Aggarwal
ISBN: 979-8-89113-305-1

Keywords: frugal innovation, entrepreneurship development, emerging markets, strategies, economic growth, sustainable development

Introduction

Frugal innovation is the process of developing low-cost, simple, and sustainable solutions to meet the needs of individuals and communities facing resource constraints. This concept is particularly relevant in emerging economies, where entrepreneurs and innovators have found creative ways to solve problems and meet the needs of their customers with limited resources. According to the OECD (2015), frugal innovation is "the development of innovative solutions to social problems, with the emphasis on affordability, accessibility and quality."

The World Bank (2016) defines frugal innovation as "the ability to deliver more value better quality, more features, or both for the same or lower cost." Prabhu and Radjou (2015) define frugal innovation as "innovation that is accessible and affordable, sustainable and profitable, and which challenges established norms to create better products, services, and business models." An example of frugal innovation is the Aravind Eye Care System, which provides high-quality eye care to low-income patients in India. By using innovative business models, technology, and operational efficiency, Aravind has been able to deliver high-quality eye care at a fraction of the cost of traditional hospitals (Govindarajan and Trimble, 2012).

Frugal innovation is becoming increasingly relevant in the global economy due to several factors. First, emerging economies are becoming more important in the global marketplace, and these economies often face resource constraints (OECD, 2015). Frugal innovation allows entrepreneurs and innovators to develop solutions that are affordable, accessible, and sustainable, enabling them to meet the needs of these markets. Second, consumers are becoming more price-sensitive and environmentally conscious, leading to a greater demand for products and services that are low-cost and sustainable (Prabhu and Radjou, 2015). Third, advances in technology and manufacturing are making it easier and more affordable to develop and distribute frugal innovations. Finally, the growth of STEM education and the increasing number of skilled workers in emerging economies are providing a strong foundation for frugal innovation (UNESCO, 2015).

STEM (science, technology, engineering, and mathematics) education is particularly important for frugal innovation, as it provides the skills and knowledge necessary to develop and implement innovative solutions. For example, many frugal innovations rely on technology to increase efficiency and reduce costs. STEM education can also help to create a culture of innovation and problem-solving, encouraging individuals and organizations to develop creative solutions to social and environmental challenges. As a result, many countries and organizations are investing in STEM education and promoting the development of frugal innovations as a means of addressing global challenges and driving economic growth (Jindal et al., 2016).

Framework of Frugal Innovation

Hossain et al., (2022) proposed a framework for frugal innovation, which includes three interrelated components: antecedents, mediators, and consequences. Antecedents refer to the factors that drive frugal innovation, such as poverty, institutional inefficiency, and scarcity of resources. Mediators are the factors that facilitate or hinder the outcomes of frugal innovation, such as innovation tasks, resource constraints, institutional constraints, and scalability constraints. Consequences refer to the economic, social, and environmental outcomes of frugal innovation, reflecting the three pillars of sustainability.

Antecedents play a crucial role in driving frugal innovation. Frugal innovation often emerges in contexts where resources are scarce, driving individuals and organizations to find creative and resource-efficient ways to meet the needs of underserved populations. Institutional inefficiency can also act as an antecedent, as bureaucratic hurdles and regulatory complexities may spur frugal innovation as a means to circumvent these constraints.

Mediators are the factors that influence the outcomes of frugal innovation. The innovation task refers to the specific problem or challenge being addressed through frugal innovation. Resource constraints are an inherent characteristic of frugal innovation, as it necessitates doing more with less. These constraints can be financial, material, or technological and require innovative approaches to achieve cost-effective solutions. Institutional constraints, such as regulatory frameworks and policies, can either facilitate or hinder frugal innovation. Finally, scalability constraints arise from the need to scale frugal innovations beyond the initial pilot or prototype stage to achieve a broader impact.

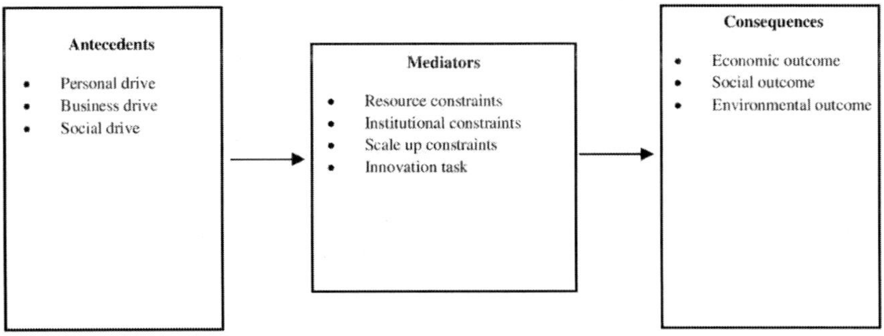

Source: Hossain et al., (2022).

Figure 1. Framework of fugal innovation.

Consequences encompass the economic, social, and environmental outcomes of frugal innovation, aligning with the three pillars of sustainability. Economically, frugal innovations can provide affordable and inclusive solutions that stimulate economic growth and improve livelihoods, especially in resource-constrained settings. Socially, frugal innovations often address the needs of marginalized communities, enhance access to essential services, and promote social inclusion. From an environmental perspective, frugal innovations tend to prioritize resource efficiency, waste reduction, and sustainability, contributing to a more sustainable and resilient future. The framework emphasizes the need to consider the interplay between these components when studying frugal innovation as shown in Figure 1.

Lessons from Successful Entrepreneurs

Case Studies of Frugal Innovation in Action

Frugal innovation is an approach that has been gaining attention for its potential to address global challenges by creating affordable and accessible solutions. One area where frugal innovation has been successful is in agriculture, where researchers at Pennsylvania State University have developed the AgroPad - a low-cost, easy-to-use testing kit that enables farmers to determine soil quality and nutrient levels using just a smartphone and a small soil sample. This technology has the potential to improve crop yields and reduce the use of fertilizers, contributing to sustainable agriculture (Qamar et al., 2020).

In healthcare, Embrace Innovations, a startup based in San Francisco, has developed an affordable infant warmer designed to prevent hypothermia in premature and low-birth-weight babies in developing countries. This low-cost solution has the potential to save countless lives in resource-constrained environments.

Frugal innovations in India, such as Tata Swach and the Aravind Eye Care System, are complemented by initiatives like the WaterScope, which further demonstrate the impact of frugal innovation in addressing critical societal challenges. Tata Swach's affordable water purifier has not only provided access to safe drinking water but has also helped prevent waterborne diseases in rural communities (Radjou and Prabhu, 2015). Similarly, the Aravind Eye Care System's efficient model has significantly reduced the cost of eye care services, enabling individuals from economically disadvantaged backgrounds to receive quality treatment (Govindarajan and Trimble, 2012).

In 2016, a group of engineering students from the Indian Institute of Technology (IIT) Delhi developed a low-cost, portable device for testing water quality in rural areas. The device called the WaterScope, consists of a microscope that attaches to a smartphone and can detect bacteria in water samples. The device is easy to use and can provide quick results, enabling communities to take action to improve water quality (Tesfaye and Fougère, 2021).

Indian healthcare startup that specializes in the development of artificial intelligence (AI)-based medical imaging solutions. Their frugal innovation involves leveraging AI algorithms to analyze medical images, such as X-rays and CT scans, for early detection and diagnosis of diseases, including tuberculosis, lung cancer, and brain hemorrhages. By automating the analysis process, Qure.ai's solutions enable faster and more accurate diagnosis, particularly in resource-constrained settings where access to radiologists may be limited. This frugal innovation has the potential to significantly improve healthcare outcomes and reduce costs, making high-quality medical imaging accessible to a larger population.

Strategies and Challenges of Frugal Innovation

Frugal innovation is not without its challenges. One of the main issues faced by innovators in this space is the need to balance cost constraints with the development of high-quality, effective solutions (George, 2016; Govindarajan and Trimble, 2012). This requires a deep understanding of local market conditions and the ability to identify and address unmet needs with limited resources. Additionally, innovators must contend with regulatory barriers and limited access to funding (Prahalad and Mashelkar, 2010). However, there are several other challenges that innovators face in frugal innovation. One major challenge is the lack of infrastructure and limited access to resources, particularly in low-income or resource-constrained settings (Halme and Korpela, 2014). Limited access to reliable electricity, connectivity, and transportation can hinder the development, distribution, and scalability of frugal innovations. Additionally, regulatory barriers, such as complex bureaucratic processes and stringent standards, can pose challenges for innovators, particularly in highly regulated industries like healthcare (Prahalad and Mashelkar, 2010). Overcoming these challenges requires creative problem-solving, collaboration with stakeholders, and engagement with policymakers to create an enabling environment for frugal innovation to thrive.

To overcome these challenges, frugal innovators often rely on several key strategies. These include leveraging existing resources and infrastructure (Halme and Korpela, 2014), developing modular and scalable solutions, and fostering partnerships and collaboration across different sectors and stakeholders (Gupta and Sodhi, 2019; Radjou and Prabhu, 2015). Innovators also focus on identifying and prioritizing the needs of underserved customers, while embracing sustainability as a core principle in the design and implementation of solutions (Nandal et al., 2019; Arora et al., 2017; Bhatti and Ventresca, 2012).

Strategies for frugal innovation also include adopting a user-centric approach and involving end-users in the innovation process. By actively engaging with the target users and understanding their needs and preferences, innovators can develop solutions that are tailored to their specific requirements (Govindarajan and Ramamurti, 2011). This approach not only ensures the relevance and effectiveness of the solutions but also increases the likelihood of adoption and acceptance in the market. User feedback and co-creation also help in iterating and refining the solutions based on real-world insights, leading to continuous improvement and innovation (Lackéus et al., 2019).

These strategies have been successfully employed in several frugal innovation initiatives, from low-cost medical devices to affordable energy solutions and beyond. Despite these successes, however, frugal innovation remains a complex and evolving field, and innovators and researchers must continue to explore new strategies and models for innovation, while also addressing the broader societal and economic challenges that underpin the need for frugal solutions.

Key Success Factors for Frugal Innovation

Frugal innovation can be successful if certain key factors are taken into consideration as shown in Table 1.

Table 1. Key success factors for frugal innovation

Key Success Factor	Description	Reference(s)
Understanding local needs	In-depth understanding of local market conditions and user needs	(George, 2016)
Leveraging existing resources and infrastructure	Utilizing available resources and infrastructure to develop solutions	(Halme and Korpela, 2014)
Modular and scalable solutions	Developing solutions that are easily replicable and scalable	(Govindarajan and Trimble, 2012)
Partnership and collaboration	Collaboration between different stakeholders, sectors, and communities	(Arora et al., 2017)
Embracing sustainability	Incorporating sustainability as a core principle in the design and implementation of solutions	(Bhatti and Ventresca, 2012)
Risk-taking and iteration	Willingness to take risks and iterate quickly in response to feedback	(Nandal et al., 2019)
Digital technology	Utilizing digital technology to create low-cost, scalable solutions	(Gupta and Sodhi, 2019)
STEM expertise	Utilizing expertise in science, technology, engineering, and mathematics to develop innovative solutions	(da Costa et al., 2021)

Source: Author's compilation.

Trends Shaping the Future of Frugal Innovation

Frugal innovation is a rapidly evolving field, and its future is being shaped by several key trends. One of these trends is the emergence of new technologies such as artificial intelligence (AI), blockchain, and the internet of things (IoT). AI, for example, is increasingly being used to develop affordable and low-cost medical devices that can improve diagnosis and treatment in underserved communities (Gupta and Sodhi, 2019). Blockchain is being used to develop frugal solutions in fields like financial services and supply chain management, where transparency and efficiency are critical (Gupta and Sodhi, 2019). The IoT is also enabling the development of frugal solutions in areas like energy and agriculture, by providing real-time data that can be used to optimize processes and reduce waste (Radjou and Prabhu, 2015).

Another trend that is shaping the future of frugal innovation is changing consumer demands. As consumers become more environmentally conscious and socially aware, there is a growing demand for products and services that are sustainable, ethical, and socially responsible. This trend is driving the development of frugal solutions that are not only affordable but also eco-friendly and socially responsible (Halme and Korpela, 2014; Schwartz and Carroll, 2003). For example, frugal innovations in the field of renewable energy are enabling access to clean energy for underserved communities, while also reducing dependence on fossil fuels (Radjou and Prabhu, 2015).

The future of frugal innovation is also being influenced by emerging STEM-based technologies in fields like materials science, engineering, biotechnology, and nanotechnology. Advances in these fields are enabling the development of new low-cost, high-performance materials and portable diagnostic tools, which have the potential to revolutionize healthcare and other industries (George, 2016; Govindarajan and Trimble, 2012). For example, frugal innovations in the field of medical devices are using nanotechnology to develop low-cost and portable diagnostic tools that can be used in remote and underserved areas (Govindarajan and Trimble, 2012).

Education and Government Policies for Promoting Frugal Innovation

Education and government policies are essential drivers for promoting frugal innovation and entrepreneurship development. STEM education provides the necessary skills and knowledge, while government policies can create an enabling environment for innovation and entrepreneurship to thrive.

STEM education is particularly crucial for promoting frugal innovation, as it provides students with the necessary skills and knowledge to develop innovative and cost-effective solutions (Saxena et al., 2020). According to Lackéus et al., (2019), entrepreneurship education programs can help aspiring entrepreneurs develop business models, conduct market research, and build financial and managerial skills. Such programs can also provide students with opportunities to develop problem-solving skills and cultivate a mindset of creativity and innovation (OECD/EU, 2019). Moreover, STEM education can promote creativity and problem-solving abilities, which are essential for entrepreneurship development (Saxena et al., 2020).

In addition to education, government policies also play a crucial role in fostering frugal innovation. In India, for instance, the government has launched several policies and initiatives to promote frugal innovation. The "Make in India" initiative aims to encourage domestic manufacturing and innovation (Govindarajan and Ramamurti, 2011). The National Innovation Foundation provides financial and technical support to innovators and entrepreneurs (Gupta, et al., 2020). Furthermore, the Atal Innovation Mission aims to foster innovation and entrepreneurship among students (Mishra and Kumar, 2019). Such government policies create an enabling environment for frugal innovation by providing financial support, tax incentives, and regulatory frameworks that encourage entrepreneurship and innovation (UNCTAD, 2018a).

Therefore, a combination of education and government policies is crucial for promoting frugal innovation and entrepreneurship development. STEM education provides students with the necessary skills and knowledge, while government policies create an enabling environment for innovation and entrepreneurship to thrive. Through education and government policies, individuals can acquire the necessary knowledge, skills, and attitudes required to become successful entrepreneurs and innovators, which can contribute to economic growth, job creation, and social welfare (Kumari et al., 2022).

Practical Guidance for Promoting Frugal Innovation and Economic Growth

Best Practices for Entrepreneurs

Entrepreneurs can adopt several best practices to promote frugal innovation and drive economic growth. One key strategy is to identify and understand the needs of the market and customers. By conducting thorough market research, entrepreneurs can gain insights into the needs and preferences of their target audience, and develop innovative solutions that meet those needs (Lackéus et al., 2019).

Another best practice is to leverage existing resources and infrastructure to minimize costs and maximize efficiency. For instance, entrepreneurs can use open-source technologies, collaborate with other organizations, and leverage shared resources to reduce the costs of innovation (Govindarajan and Ramamurti, 2011).

Entrepreneurs can also adopt a flexible and agile approach to innovation. By constantly testing and refining their products and services based on customer feedback, entrepreneurs can develop solutions that better meet the needs of the market and are more likely to succeed in the long run (Lackéus et al., 2019).

Moreover, entrepreneurs can seek partnerships with academic institutions, government agencies, and non-profit organizations to access funding, mentorship, and technical expertise. Such partnerships can also help entrepreneurs to access new markets, build networks, and develop innovative solutions that address social and environmental challenges (Schwartz and Carroll, 2003).

Finally, entrepreneurs can prioritize sustainability and social responsibility in their innovation efforts. By developing solutions that have a positive impact on society and the environment, entrepreneurs can attract customers, investors, and partners who value sustainability and social responsibility (Govindarajan and Trimble, 2012).

In summary, entrepreneurs can adopt several best practices to promote frugal innovation and drive economic growth. By understanding customer needs, leveraging existing resources, adopting a flexible approach to innovation, seeking partnerships, and prioritizing sustainability and social responsibility, entrepreneurs can develop innovative solutions that create value for society and contribute to economic growth.

Policy Recommendations for Policymakers

Table 2 summarizes the policy recommendations for promoting frugal innovation and economic growth.

Table 2. Policy recommendations for promoting frugal innovation and economic growth

Policy Recommendations	References
Create an enabling policy environment	"Governments can create an enabling environment for frugal innovation by providing financial support, tax incentives, and regulatory frameworks that encourage entrepreneurship and innovation" (UNCTAD, 2018b).
Promote innovation in public services and procurement	"Governments can also promote innovation in public services and procurement, which can help to stimulate demand for frugal innovation and create new market opportunities for entrepreneurs" (OECD, 2019a). Additionally, policymakers can organize innovation challenges and competitions to identify and recognize innovative ideas and solutions. By creating an environment that celebrates and rewards innovation, policymakers can incentivize frugal innovation and drive economic growth (Gupta et al., 2020).
Support incubation centers and technology parks	"Governments can create incubation centers, provide seed capital, and establish technology parks to support the development and commercialization of frugal innovations" (Lakshmi and Gautam, 2019).
Develop human capital through education and training	"Entrepreneurship education programs can help aspiring entrepreneurs to develop business models, conduct market research, and build financial and managerial skills" (Lackéus et al., 2019).
Strengthen intellectual property rights	"Strong intellectual property rights protection can provide incentives for innovation and investment in research and development" (UNCTAD, 2018b).
Encourage collaboration and knowledge-sharing	Policymakers can facilitate collaboration between different stakeholders, such as entrepreneurs, academia, industry experts, and non-profit organizations. By creating platforms and networks for knowledge sharing and collaboration, policymakers can foster an environment where ideas, expertise, and resources can be shared, leading to innovative solutions and economic growth. This can be done through the establishment of innovation hubs, incubators, and accelerators that bring together diverse stakeholders and provide them with the necessary support and resources to thrive (Lackéus et al., 2019).

Table 2. (Continued)

Policy Recommendations	References
Invest in education and skill development	Investing in education and skill development not only benefits individuals but also enhances the overall innovation ecosystem and supports sustainable economic development (Schwartz and Carroll, 2003). By investing in human capital, policymakers can create a strong foundation for frugal innovation and economic growth, enabling individuals to actively participate in the innovation process and contribute to societal progress.
Enhance access to finance	Policymakers can collaborate with financial institutions to design innovative financial products that support frugal innovation projects. By addressing the financial barriers faced by frugal innovators, policymakers can unlock their potential and contribute to economic growth (Kaplinsky et al., 2013).

Source: Author's compilation.

Conclusion

In conclusion, frugal innovation has emerged as a valuable approach for entrepreneurs seeking to create innovative products, services, or business models with limited resources. Through the examination of case studies of successful entrepreneurs, valuable insights have been gathered that highlight the key success factors and challenges involved in adopting such an approach. Its relevance has been particularly highlighted in the education sector, with STEM education playing a crucial role in promoting frugal innovation and preparing the next generation of entrepreneurs. Additionally, emerging technologies, changing consumer demands, and new business models are shaping the future of frugal innovation. It is important for policymakers and educators to recognize the importance of frugal innovation in promoting entrepreneurship development and sustainable economic growth, and to provide guidance and support for entrepreneurs seeking to adopt this approach. Practical guidance for promoting frugal innovation and economic growth has been discussed for both entrepreneurs and policymakers to follow. Frugal innovation is an important concept that will continue to shape the future of innovation, and individuals and organizations must stay abreast of this trend to remain competitive in the global economy.

Future Research Directions

Frugal innovation has been gaining momentum in recent years as a way to address the challenges of affordability, sustainability, and inclusivity in the global market. There is a growing recognition that frugal innovation can have a significant impact on economic growth and development, particularly in emerging economies. Many successful entrepreneurs have demonstrated that frugal innovation can lead to profitable business models while addressing the needs of low-income consumers.

In the context of education, there is a need for frugal innovation to address the challenges of providing quality education at an affordable cost. STEM education, in particular, can play a crucial role in promoting frugal innovation in education. STEM education can provide the necessary skills and knowledge to students to develop innovative solutions to address the challenges faced by low-income communities. By promoting STEM education, policymakers can foster a culture of innovation and entrepreneurship, which can lead to sustainable economic growth. One such example is that of Pratham Books, a non-profit organization based in India that focuses on providing affordable books in local languages for children. The organization uses a frugal innovation approach to publish high-quality books in multiple languages at a low cost, making them accessible to children from diverse backgrounds.

In addition, several initiatives use frugal innovation in the field of STEM education. For instance, the Science Education Programme in Rural Areas (SEPRA) in Ghana uses a low-cost approach to provide science education to children in rural areas. The program uses locally available resources to create science kits and conducts experiments that are relevant to the local context. These examples demonstrate the potential of frugal innovation in promoting STEM education and creating opportunities for individuals from diverse socio-economic backgrounds.

Looking to the future, there is a need for more research and investment in frugal innovation. Policymakers need to take proactive steps to create an enabling environment that promotes frugal innovation. This includes investing in research and development, providing access to finance and technology, and promoting collaboration between entrepreneurs, researchers, and policymakers. Additionally, there is a need to develop new business models and approaches that can better serve low-income consumers. With a continued focus on frugal innovation and STEM education, it is possible to create a more inclusive and sustainable future for all.

References

Arora, P., Pruthi, S., and Dwivedi, Y. K. (2017). Co-creation of value in frugal innovations. *Journal of Business Research*, 79, 1-13.

Bhatti, Y., and Ventresca, M. (2012). Sustainability and 'Frugal Innovation': Evidence from India. *Journal of Management Inquiry,* 21(2), 192-208.

da Costa Nogami, V. K., & Veloso, A. R. (2021). Innovation in the subsistence marketplace: an analysis considering multiple concepts and approaches. *Innovation & Management Review*, *18*(1), 2-16.

George, G. (2016). Frugal Innovation and the 'Quiet' Emergence of the Global South. *California Management Review,* 59(4), 6-22.

Govindarajan, V., and Ramamurti, R. (2011). India's Frugal Innovations. *Harvard Business Review,* 89(8), 1-9.

Govindarajan, V., and Trimble, C. (2012). Frugal Innovation: A Game-Changing Business Model. *Harvard Business Review*, 90(5), 82-90.

Gupta, A., Srivastava, K., and Sushil. (2020). Towards Creating an Enabling Ecosystem for Frugal Innovations in Emerging Economies. *Journal of Innovation and Entrepreneurship,* 9(1), 1-22.

Gupta, M., and Sodhi, D. (2019). Artificial Intelligence in Healthcare: Past, Present, and Future. *International Journal of Healthcare Management*, 12(2), 89-94.

Halme, M., & Korpela, M. (2014). Responsible innovation toward sustainable development in small and medium-sized enterprises: A resource perspective. *Business Strategy and the Environment*, *23*(8), 547-566.

Hossain, M., Agarwal, N., Bhatti, Y., & Levänen, J. (2022). Frugal innovation: Antecedents, mediators, and consequences. *Creativity and Innovation Management*, *31*(3), 521-540.

Jindal, P., & Tej, F. (2016). Being an educator in private vs. public school system: A comparative study on job contentment. *IRA International Journal of Education and Multidisciplinary Studies*, *3*(3), 309-319.

Kumari, S., Jindal, P., & Chand, P. K. (2022). Social Media Networking: As Facilitator of Acculturation. *ECS Transactions*, *107*(1), 9381.

Lackéus, M., Ellegård, K., and Klinton, M. (2019). Entrepreneurship Education: A Systematic Review of the Methods in Entrepreneurship Education Programs and Their Impact on the Students. *Journal of Entrepreneurship Education*, 22(2), 1-33.

Mishra, R., Ramesh, and Ashish Kumar. (2019). A study of the impact of Atal Innovation Mission on promoting innovation and entrepreneurship among students in India. *Journal of Entrepreneurship Education,* 22(3), 1-11.

Modgil, S., Gupta, S., & Bhushan, B. (2020). Building a living economy through modern information decision support systems and UN sustainable development goals. *Production Planning & Control*, *31*(11-12), 967-987.

Nandal, V., Kumar, R., & Singh, S. K. (2019). Barriers identification and analysis of solar power implementation in Indian thermal power plants: An Interpretative Structural Modeling approach. *Renewable and Sustainable Energy Reviews*, 114, 109330.

OECD. (2015). Innovation Policies for Inculsive Growth. *Paris:* OECD Publishing. Retrieved from https://www.oecd.org/sti/innovation-policies-for-inclusive-growth-9789264229488-en.htm on May 22, 2015.

OECD. (2019a). Promoting Frugal Innovation for Inclusive Development. OECD Science, *Technology and Innovation Policy Papers,* (52), 1-57.

OECD/EU. (2019). The Future of Education and Skills: Education 2030. *OECD Education Working Papers,* (201), 1-146.

Prabhu, J., and Radjou, N. (2015). Frugal Innovation: A New Business Paradigm. In M. Dodgson, D. M. Gann, and N. Phillips (Eds.), *The Oxford Handbook of Management: Oxford Handbook of Innovation Management*, Oxford: Oxford University Press, 282-326.

Prahalad, C. K., and Mashelkar, R. A. (2010). Innovation's Holy Grail. *Harvard Business Review,* 88(7/8), 132-141.

Qamar, S. R., Ali, S., Hussain, M., Hussain, A., and Hussain, T. (2020). AgroPad: A low-cost sensing technology for precision agriculture. *IEEE Sensors Journal,* 20(13), 7285-7292.

Radjou, N., and Prabhu, J. (2015). Frugal innovation: How to do more with less. *The Economist.* Retrieved from https://www.economist.com/special-report/2015/10/22/how-to-do-more-with-less on *October 22, 2015.*

Saxena, R., Sharma, P., and Kaur, P. (2020). Frugal Innovation in Education: Role of STEM Education. *International Journal of Emerging Technologies in Learning,* 15(3), 67-77.

Schwartz, M. S., and Carroll, A. B. (2003). Corporate Social Responsibility: A Three-Domain Approach. *Business Ethics Quarterly*, 13(4), 503-530.

Tesfaye, L. A., & Fougère, M. (2021). Frugal innovation hijacked: The co-optive power of co-creation. *Journal of Business Ethics*, 1-16.

The World Bank. (2016). Frugal Innovation: Doing More with Less. Washington, D.C.: World Bank Group.

UNCTAD. (2018a). Policies to Foster Frugal Innovation for Development. UNCTAD/DTL/STICT/2018/5, United Nations Conference on Trade and Development.

UNCTAD. (2018b). Technology and Innovation Report 2018: Harnessing Frontier Technologies for Sustainable Development. United Nations Conference on Trade and Development. Retrieved from https://unctad.org/en/PublicationsLibrary/tir2018_en.pdf.

UNESCO. (2015). Engineering: Issues, Challenges and Opportunities for Development. Paris: UNESCO Publishing.

Zeschky, M. B., Winterhalter, S., & Gassmann, O. (2014). From cost to frugal and reverse innovation: Mapping the field and implications for global competitiveness. *Research-Technology Management*, *57*(4), 20-27.

Chapter 14

STEM Employees: The Role of Co-Worker Support, Perceived Insider Status and Organisation Citizenship Behaviour in Achieving Employees' Retention

Koshish Jyoti[1]
Garima Mathur[2,*], PhD
and Yogesh Upadhyay[1], PhD
[1]School of Studies in Management, Jiwaji University, Gwalior, India
[2]Prestige Institute of Management & Research, Gwalior, India

Abstract

The proposed chapter deals with analyzing the effect of organizational citizenship behaviour and perceived insider status on employee retention and finding out the relationship between co-worker support, perceived insider status, and organizational citizenship behaviour. The study is based on a survey 300 of employees working in STEM fields in Madhya Pradesh in India. Regression analysis and correlation were used to test the hypothesis in our research model using SPSS version 2018. This is the study that focuses on the role of co-worker support, perceived insider status, and organizational citizenship behaviour concerning STEM employee retention. The finding depicts that Co-worker support has a positive relationship with perceived insider status and organizational citizenship behaviour and both perceived insider status and OCB has a positive effect on employees' retention in the organization. Organizations must try to focus on co-worker support, perceived insider

* Corresponding Author's Email: garimanmathur@gmail.com.

In: STEM
Editors: Priya Jindal, Amandeep Singh and Rashmi Aggarwal
ISBN: 979-8-89113-305-1

status, and organizational citizenship behaviour aspects and also concentrate on other factors as well such as recognition, and rewards. They should try to provide a favorable workplace environment to the employees to retain themselves in the organization.

Keywords: organization citizenship behaviour, STEM employees, perceived insider status, employee retention, worker support

Introduction

Managers struggle to retain employees because they are an organization's most important, valuable, and productive asset. High employee turnover can be avoided by adopting a comprehensive retention strategy to deal with such problems.

This study will critically evaluate employee retention research, with the active involvement of perceived insider status and organizational citizenship behaviour. To bring the disparate knowledge together and present the literature review in an organized way, leading to an exploratory study, several secondary sources have also been examined along with the primary source. Due to technological advancements, increased competition, and increased globalization, employees have taken on a greater significance in modern businesses. The literature review presents both conventional and novel approaches to employee retention. Furthermore, greater emphasis will be placed on elements like the need for professional growth and skill development, the capacity to exercise control over decision-making, and the ever-increasing goal of fulfilling employees. The long-term success and viability of any business depends on its ability to hold on to its most valuable assets—its employees. If you want satisfied co-workers and reporting staff, smooth succession planning, etc., you need to be able to hold on to your best employees. The challenges are more in the case of employees from computer backgrounds or individuals like scientists, mathematicians, technocrats, etc., popularly known as STEM employees.

The practice of praising and rewarding workers for remaining with the company for many years is known as employee retention. During this time, employees are strongly encouraged to remain with the firm for as long as possible or until the endeavor is completed. According to Maertz and Campion (1998), relatively less research on employee retention has concentrated specifically on how an employee chooses to remain with an organization and what determines this attachment. Both retention and quitting processes should be researched. According to Zineldin (2006), retention is "an obligation to do business or exchange with a particular company continuously." Employees who are content with their jobs, (Denton, 2000) are more committed to their work and constantly make efforts to increase organizational customer satisfaction. Retention is defined as "customer liking, identification, commitment, trust, readiness to recommend, and repurchase intentions," with the first four being emotional-cognitive retention constructs and the last two being behavioural intentions, (Stauss et al., 2001). Panoch (2001) advanced the idea that good employees are getting harder and harder to find, organizations today take great care to keep them around. Due to their importance to an organization's success, organizational citizenship behaviors have gained significant attention in academia over the past few decades Citizenship in the Workplace (Roberts et al., 1998).

The term 'organizational citizenship behaviors' (OCBs) was coined by (Smith et al., 1983) to describe any actions taken by employees outside the scope of their job descriptions but still beneficial to the organization. Barnard (1938) suggested that workers' free will and deliberate actions matter greatly to their employers is a key foundation for this argument. According to Barnard (1968), organizations are cooperative systems, the success of which depends on the cooperation of its members. Despite evidence showing the impact of OCB on employee retention (Paille' and Grima, 2011; Lam et al., 2009; Harrison et al., 2006; Chen et al., 1998; MacKenzie et al., 1998) some studies have argued that further study is needed to better understand the relationships between OCBs and variables involved in the decision of employee retention. Human resource management (HRM) professionals and academics alike have recently begun focusing on the correlation between HRM policies and procedures and turnover rates. Employees are often seen as irreplaceable resources by their employers (Redman and Wilkinson, 2001), making this a pressing issue. Their planned departure will almost certainly have an impact on how efficiently and effectively the company functions. Talented and skilled workers are hard to come by in today's job market.

Co-Worker Support

Co-worker support refers to when co-workers help each other when they genuinely need help with practical knowledge, work, or active encouragement in the organization (Zhou and George, 2001). According to Albar et al.,

(2005), co-worker support is effective support in terms of emotional context overwork stress. It also stated that co-worker support has a positive effect on prospective employees' inclusive work environment (Mcintosh, 1991). Karatepe et al., (2010) unequivocally discovered that co-worker support significantly reduces stress levels and resolves work-related issues. According to Baumeister and Leary (1995), the direct relationship between co-worker support is derived from social exchange theory, which indicates that employees who interact with each other create a good feeling for each other, and the reciprocity theory of Gouldner, which properly represents that an individual's inherent tendency that what you give in return of this you receive the same. Allen and Shanock (2013) concluded from their necessary research that the co-worker's support factor for employee turnover has a significant impact on organizations. According to Tews et al., (2013), "help with noncore responsibilities is more valued because it may be perceived as a friendly helping hand, whereas help with core responsibilities may be perceived as implying a lack of competence."

Perceived Insider Status

It is defined as an insider perception as "an adequate representation of the degree to which an individual employee of a specific company instantly perceives himself or herself as an insider" (Stamper and Masterson, 2002). PIS occurs when an employee feels like they belong to the company like a member of the family pioneering research. PIS by (Dai and Chen, 2015; Stamper and Masterson, 2002) also shed additional light on this. People want to connect with others and feel like they belong, so they actively seek out meaningful relationships.

"Achieving an insider status gives certain members of the group access to more resources and contributes more to group-derived self-esteem" (Gómez, 2012). When employees feel valued and appreciated by their employer, they are more likely to go above and beyond in their work for the company, viewing themselves as 'ultimate insiders'. In contrast, workplace incivility harms PIS if employees are subjected to a high volume of disrespectful behaviour from their superiors and co-workers (Guo et al., 2020). Employees', "perceived insider status" (PIS) contributes to their happiness, success, and commitment to the company (Nishii and Mayer, 2009; Barak and Levin, 2002).

Benefits of PIS include "favoring acceptation of responsibilities, respect of norms, performance, professional satisfaction, and retention, economic well-being in the workplace, creativity, and organizational citizenship behaviours" (Tang et al., 2023).

Organization Citizenship Behaviour

The first empirical study of organizational citizenship behavior was conducted by (Bateman and Organ, 1983). Since then, many other researchers have contributed to our understanding of this developing field. Borman and Motowidlo (1997); Munene (1995); George and Bettenhausen (1990); Bateman and Organ (1983) explain "discretionary behaviors on the part of an employee that directly promote the effective functioning of an organization, independent of an employee's objective." Organ (2014) focused on the term OCB a 'willingness to cooperate', which was brought to light by the field of industrial and organizational psychology.

Employee Retention

The goal of employee retention is to ensure that employees stay with an organization for as long as possible. "Technique adopted by businesses to maintain an effective workforce and at the same time meet operational requirements," defined by (Mita et al., 2014). To stay with the company for as long as possible or until the project is finished is the goal of this process, as defined by (Das and Baruah, 2013). According to Workforce Planning for Wisconsin State Government (2005), "employee retention is a systematic effort to create and foster an environment that encourages employees to remain employed by having policies and practices in place that address their diverse needs." The success of an organization in keeping its employees depends on many factors, including its vision, mission, values, and policies.

Objectives

- To find out the relationship between co-worker's support with perceived insider status and organizational citizenship behaviour.
- To find out the impact of perceived insider status and organizational citizenship behaviour.

Literature Review

Co-Worker Support and Perceived Insider Status

While there have been no empirical studies of PIS in conjunction with social support in the workplace, (Baumeister and Leary's, 1995) theory of belongingness is related to PIS because it, too, reflects an employee's acceptance or rejection by their peers. Employees' reactions to favors from co-workers and the organization as a whole, such as morale willingness to speak up organization identification, and turnover intention (Sarah et al., 2021) can be explained by the support they receive from their peers.

When you have the backing of your respected co-workers, you'll feel more at home in the company and more invested in its success. An employee's sense of belonging to the company increases when he or she believes the company supports them. Davila and Garcia (2012) argue that people suffer when they are excluded because doing so undermines their basic human desire to feel accepted by others. These focused requirements actively improve patient security, outcomes, and viability. For thousands of years, many different cultures have willingly engaged in the practice of social exclusion. Its effectiveness is demonstrated here, but so is the aesthetic preference of humans for social order and relationships. Since exclusion is probably here to stay, it's important to look into ways to mitigate its impact on businesses.

Co-Worker Support and Organizational Citizenship Behaviour

Employees can explicitly direct their citizenship behavior toward peers, as suggested by (Malatesta and Byrne, 1997). To put it another way, co-workers of roughly equal status-including teammates, team members, and workgroup members can be thought of as special beneficiaries of individually directed citizenship behavior. Numerous labels, including OCBI, have been used to describe this behavior in earlier studies that focused on peers as recipients of citizenship. Social exchange theory (Blau, 1964) and reciprocity theory (Gouldner, 1960) form the theoretical foundations for the connection between perceived support and employee behavior.

Interpersonal exchanges are strengthened by acts that demonstrate care, solidarity, and respect for one another, all of which are indicative of a long-term commitment to one another (Burleson and Samter, 1990). This is in line with findings from studies of leader-member exchange (Wayne et al., 1997) which show that citizenship behavior is prevalent in mutually beneficial relationships. Because people want to reciprocate the goodwill shown to them, we anticipate that person-centered OCB emerge as a result. Co-worker support, trust, perspective-taking, and empathic concern were all associated with person-focused OCB, as reported by (Settoon and Mossholder, 2002). These findings line up with what has been found before.

Co-worker support was found to be correlated with person-focused organizational citizenship behavior (OCB), which suggests that co-worker support may create a feeling of indebtedness among individuals as well as a concomitant desire to reciprocate positive actions directed toward them. This concept is consistent with the distinction between person-focused and task-focused OCB, the latter of which involves more directly assisting co-workers with difficult assignments, heavy workloads, and so on. Recent research by Susskind et al., (2003) found that when restaurant workers feel like their co-workers have their backs, they provide better service to customers. Focus on contemporaneous disciples Allen (2006) discovered that co-worker support is significantly and positively correlated with organizational citizenship behavior, indicating that organizationally directed citizenship and co-worker-directed citizenship are distinct constructs.

Perceived Insider Status and Employee Retention

There was a positive correlation between organizational identification and job satisfaction and a negative correlation between organizational identification and turnover intentions, but this was only discovered through post hoc analysis. Previous studies found conflicting results when examining the links between civic virtue and the decision to give up smoking. Whereas Paille and Grima (2011) did not find any associations between civic virtue and intention to quit. Jindal and Tej (2016) found that perceived insider status was a significant predictor of job satisfaction and turnover intentions even after controlling for psychological ownership and organizational identification. Perceived insider status is hypothesized to have a positive effect on job satisfaction and a negative effect on the intention to leave one's current position, following the aforementioned theoretical rationale. However, insider status has significant positive effects (i.e., decreased turnover intentions and increased role performance) for even non-enhanced jobs. There are outlined several factors that are essential to keeping older workers in the workforce (Raub, 2018). A sense of community, fair treatment from superiors, regular performance reviews, and praise all play important roles in retaining older workers. Whether or not retirees choose to stay with their company is greatly influenced by how much they feel like they belong there.

Employees' responsibilities to the company as a result of the investment's success are an integral part of the social exchange cycle. There is an obligation to be present as a member of a group. Therefore, staying with an organization is an obvious result of being a member of that organization. It was suggested that insider status and turnover intentions be studied together (Stamper and Masterson, 2002).

It is proposed that an absence of social integration or inclusion in the workplace is a contributing factor to employee turnover (Kumari et al., 2022). As a result, we hypothesized that older workers who saw themselves as insiders within their organization would be more committed to staying there than those who saw themselves as outsiders.

Organizational Citizenship Behaviour and Employee Retention

As a result of the positive work environment and climate created by OCBs like altruism and sportsmanship, organizations that actively encourage these traits among their employees are more likely to attract and keep the most talented workers (Shanker et al., 2021). The association between OCB and the intention to leave an organization has been the subject of investigation in several previous works (Podsakoff et al., 2009). In addition, discovered that OCB intensities represent employees' genuine propensities, whether they want to be actively engaged with their organization or would prefer to stay away from it. Many studies (Krishnan and Singh, 2010; Mossholder et al., 2005) have used a global measure of OCB to examine its effect on employee retention (intended and actual turnover). Some study focuses primarily on an effort by examining whether forms of OCB increase or decrease the intention to search and subsequently affect the intention to leave the organization (Thau et al., 2004; Hui et al., 1999). This is because previous research has been conducted to explain how perceived job alternatives affect OCB and how it is linked to the intention to leave the organization through the intention to search. Previous research discovered a negative relationship between forms of OCB and the intention to leave the organization (Paille' and Grima, 2011).

Paillé et al., (2013) found that OCB is positively related to turnover intention, which was a surprising finding, and they suggested that more research is needed to fully comprehend the connection between OCB and turnover intention. Similarly, Lau et al., (2016) discovered that self-rated OCB positively increased turnover intention, implying that self-rated OCB did not affect turnover intention. Furthermore, Saraih et al., (2017) discovered that OCB has no relationship with turnover intention. They concluded that retention-oriented compensation and formalized training have a positive correlation with OCB engagement, but a negative correlation with OCBs' intentions to leave the organization. Employees who have a stronger sense of belonging to their company are less likely to consider quitting.

Research Methodology

The Study and Population

The study was empirical and completed with a field survey. All of the employees from various backgrounds made up the population. The sampling component was a single respondent from a background from either of the streams such as Science, Mathematics, Engineering, or other technological backgrounds. To conduct the research study, a purposive non-random sampling technique was employed. The respondents were contacted either personally or through referral networks. They were confirmed for their willingness to respond and only respondents representing the STEM field were considered. Since these employees are busy the information was gathered from 250 respondents, We have considered only 230 responses as the remaining either did not respond completely or were outliers.

Measures

Co-Worker Support

The Social Support scale designed by O'Driscoll (2000) was used to measure co-worker support with alpha .834.

Perceived Insider Status

Responses on a point Likert type scale ranging from Perceived insider status were measured using Stamper and Masterson's (2003), 6-item measure and all the items of it having Cronbach alpha value is .854 on a point Likert type scale ranging from strongly disagree to strongly agree.

Organizational Citizenship Behaviour

Organizational citizenship behaviour was measured using Spector et al., (2010) of 10 items having a Cronbach alpha value of .803 on a Likert-type scale ranging from strongly disagree to strongly agree.

Employee Retention

The employee retention seven factors scale identified by Walker (2001) was used having a Cronbach alpha value of .790 on a Likert-type scale ranging from strongly disagree to strongly agree.

Based on the proposed model as depicted in Figure 1 the following hypotheses are framed.

Hypotheses

H1: Co-worker support has a positive relationship with perceived insider status.
H2: Co-worker support has a positive relationship with organizational citizenship behaviour.
H3: Perceived insider status has a positive significant impact on employee retention.
H4: OCB has a positive significant impact on employee retention.

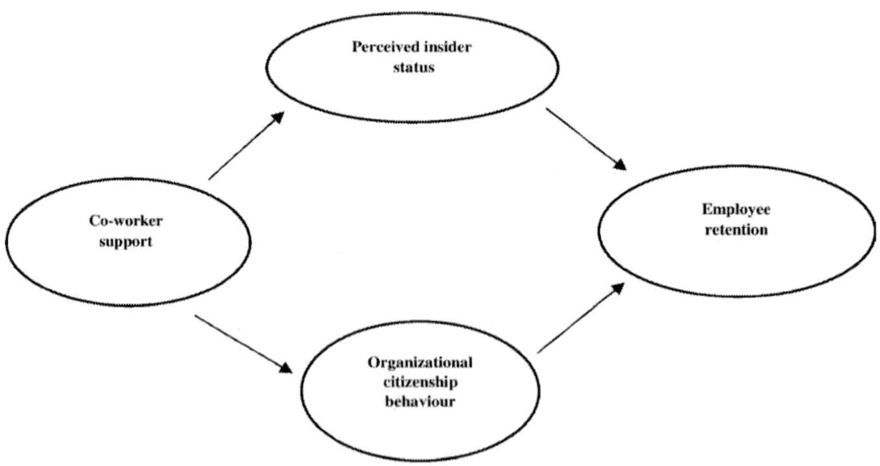

Source: Author's compilation.

Figure 1. Proposed model.

Result Interpretation

Table 1. Reliability

Variables	No. of Items	Cronbach's Alpha Based on Standardized Items	Cronbach's Alpha
Co-workers support	4	.802	.802
Perceived insider status	6	.865	.854
Organizational citizenship behavior	10	.824	.803
Employee retention	7	.751	.790

Source: Author's compilation.

By analyzing Cronbach's alpha, the reliability scores of the collected data were initially assessed in Table 1. According to Nunnally (1978), the constructs "Co-worker support," "Organizational citizenship behaviour," "Perceived Insider Status," and "Employee retention," all have good Cronbach's alpha values above 0.7, which indicated that the data collected would be consistently consistent across the board.

Table 2. ANOVA

ANOVA						
Model		Sum of Squares	Df	Mean Square	F	Sig.
1	Regression	58.372	1	58.372	80.326	0.000[a]
	Residual	149.378	219	0.682		
	Total	207.750	220			
a. Predictors: (Constant), Perceived insider status						
b. Dependent Variable: Employee retention						

Source: Author's compilation.

Table 2 shows that a model used for regression has a good fit as indicated by F-value 80 % which is significant at 0.000% level of significance indicating high predictability of the model.

The coefficient shown in Table 3 indicated that Perceived insider status was found to have a positive and significant effect on Employee retention having a β value is .526 (significant as $p < 0.005$). Hence our hypothesis (H3) is supported.

Y = a + bx + E
Y = 1.218 + .579X + E
X = (Perceived insider status) Independent variable
Y = (Employee retention) dependent variable
E = Error

Table 3. Coefficient

Coefficients						
Model		Unstandardized Coefficients		Standardized Coefficients	T	Sig.
		B	Std. Error	Beta		
1	(Constant)	1.218	.220		5.536	.000
	PIS	.579	.063	.526	0.919	.000
a. Dependent Variable: Employee retention						

Source: Author's compilation.

Table 4. Model summary

Model Summary				
Model	R	R Square	Adjusted R Square	Std. Error of the Estimate
1	.654[a]	.625	.611	1.66760
a. Predictors: (Constant), Organizational citizenship behavior				
b. Dependent Variable: Employee retention				

Source: Author's compilation.

Table 5. ANOVA

ANOVA						
Model		Sum of Squares	Df	Mean Square	F	Sig.
1	Regression	80.372	1	80.372	80.326	0.000[a]
	Residual	142.378	219	.754		
	Total	222.75	220			
a. Predictors: (Constant), Organizational citizenship behavior						
b. Dependent Variable: Employee retention						

Source: Author's compilation.

The value of R square is 0.625 in Table 4 which shows that perceived insider status explains 62.5% variance towards the dependent variable which is employee retention.

The model used for regression as shown in Table 5 has a good fit as it is indicated by F-value 80 % which is significant at 0.000% level of significance indicating high predictability of the model.

Table 6. Coefficient

Coefficients						
Model		Unstandardized Coefficients		Standardized Coefficients	T	Sig.
		B	Std. Error	Beta		
1	(Constant)	1.345	.237		5.675	.001
	Organizational citizenship behavior	.459	.096	.654	4.781	.000
a. Dependent Variable: Employee retention						

Source: Author's compilation.

The coefficient shown in Table 6 indicated that Perceived insider status was found to have a positive and significant effect on Employee retention having a β value of .654 (significant as $p < 0.005$). Hence our hypothesis (H4) is supported.

$Y = a + bx + E$
$Y = 1.345 + .459X + E$
X = (Organizational citizenship behavior) Independent variable
Y = (Employee retention) dependent variable
E = Error

Correlation Analysis

Table 7 Correlation Analysis showed that Co-worker support has positively correlated with perceived insider status as well as with organizational citizenship behaviour as they have positive value so our proposed hypotheses named H1 and H2 are supported.

Table 7. Correlation analysis

Variables	Mean	SD	1	2	3	4
Co-workers support	23.4	.24	1			
Perceived insider status	65.20	.421	.231	1		
Organizational citizenship behaviour	56.21	.25	.354	.245	1	
Employee retention	42.21	.36	.241	.214	.208	1

Source: Author's compilation.

Table 8. Summary table

Hypotheses	Result reported
H1: Co-worker support has a positive relationship with perceived insider status.	Supported
H2: Co-worker support has a positive relationship with organizational citizenship behaviour.	Supported
H3: Perceived insider status has a positive significant impact on employee retention.	Supported
H4: OCB has a positive significant impact on employee retention.	Supported

Source: Author's compilation.

Table 8 shows that all the framed hypotheses are supported by the pieces of literature done previously and have positive relations between them.

Conclusion

Though commitment issues in terms of retaining with the same organization is a relatively tough task yet not impossible. With due support and understanding it is possible to create a good environment for the committed workforce. The employees from STEM can be managed properly with a deep routed feeling of perceived insider status and OCB.

Through our study we have observed that our all proposed hypothesis is supported as Co-worker support has a positive relationship with organizational citizenship behaviour and perceived insider status when employees need help in the workplace during work commitments and get help from their subordinates, supervisors tend to have a more positive outcome in the form of citizenship behavior and sense of belongingness towards achieving organization goal. Perceived insider status and organizational citizenship behaviour have a positive significant impact on employee retention. Employees will stay in the organization when they are treated as insiders to the company. Research findings by Jewell (1985) and Locke (1976) have clearly stated that people, who are satisfied with their jobs, tend to stay in them longer, i.e., lower turnover, and be less absent.

Supporting the view Noah (2008) found in his research that employee involvement in decision-making helps in creating a sense of belongingness among the employees, which helps in creating a good congenial working environment and contributes towards building a good employer-employee relationship. There is still a lot of room for more exploration in the field of employee retention and by considering factors like compensation practices, leadership and supervision, career planning and development, alternative work schedules, working conditions, and fleecy perks, even though this paper did its best to reveal the various research works have done and the contributions forward by various researchers in the area Employees has a strong sense of commitment tend to not leave the organization. so the organization must look over smooth policies towards the employees to make them comfortable, satisfy their needs, and understand them as family members of the organization.

Limitations

- The research has several limitations that happen naturally. Nevertheless, several works were done in this studyto minimize the problems. There are the lists of limitations that this study encountered:
- The sample of this study is very small.
- Another limitation is that the data comes only from the Gwalior area, so it can't be generalized to other places. Few people filled out the survey if the questions are any indication. The survey is vulnerable to the respondents' preconceived notions and opinions. Therefore, absolute precision cannot be guaranteed.
- The time frame in which the research is conducted is so short that it does not allow for a more comprehensive examination of the topic.

Suggestions

It is suggested that a proportional stratified random sample be used in future research to compare multiple public sectors with a larger sample size. More research has to be done by future researchers on these variables, also research on different fields of study has these variables.

References

Allen, D. G., & Shanock, L. R. (2013). Perceived organizational support and embeddedness as key mechanisms connecting socialization tactics to commitment and turnover among new employees. *Journal of Organizational Behavior*, *34*(3), 350-369.

Albar Marín, M., & García-Ramírez, M. (2005). Social support and emotional exhaustion among hospital nursing staff. *The European journal of psychiatry*, *19*(2), 96-106.

Borman, W. C., & Motowidlo, S. J. (1997). Task performance and contextual performance: The meaning for personnel selection research. *Human performance*, *10*(2), 99-109.

Barnard, C. I. (1938). *The functions of the executive*. Cambridge, MA: Harvard University.

Barnard, C. I. (1968). *The functions of the executive* (Vol. 11). Harvard university press.

Baumeister, R. F., & Leary, M. R. (1995). The need to belong: desire for interpersonal attachments as a fundamental human motivation. *Psychological bulletin*, *117*(3), 497.

Barak, M. E. M., & Levin, A. (2002). Outside of the corporate mainstream and excluded from the work community: A study of diversity, job satisfaction and well-being. *Community, Work & Family*, *5*(2), 133-157.

Bateman, T. S., & Organ, D. W. (1983). Job satisfaction and the good soldier: The relationship between affect and employee "citizenship." *Academy of management Journal*, *26*(4), 587-595.

Burleson, B. R., & Samter, W. (1990). Effects of cognitive complexity on the perceived importance of communication skills in friends. *Communication Research*, *17*(2), 165-182.

Baumeister, R. F., & Leary, M. R. (1995). The need to belong: desire for interpersonal attachments as a fundamental human motivation. *Psychological bulletin*, *117*(3), 497.

Blau, P. M. (1968). Social exchange. *International encyclopedia of the social sciences*, *7*(4), 452-457.

Coyne, I., & Ong, T. (2007). Organizational citizenship behaviour and turnover intention: A cross-cultural study. *The International Journal of Human Resource Management*, *18*(6), 1085-1097.

Chen, X. P., Hui, C., & Sego, D. J. (1998). The role of organizational citizenship behavior in turnover: Conceptualization and preliminary tests of key hypotheses. *Journal of applied psychology*, *83*(6), 922.

Dávila, M. C., & García, G. J. (2012). Organizational identification and commitment: correlates of sense of belonging and affective commitment. *The Spanish journal of psychology*, *15*(1), 244-255.

Das, B. L., & Baruah, M. (2013). Employee retention: A review of literature. *Journal of business and management*, *14*(2), 8-16.

Denton, J. W. (2000). Using Web-based projects in a systems design and development course. *Journal of Computer Information Systems*, *40*(3), 84-87.

Dai, L., & Chen, Y. (2015). A systematic review of perceived insider status. *Journal of Human Resource and Sustainability Studies*, *3*(02), 66.

Gomez, S. (2012). *Offering Hope from the Inside Out One Stroke at a Time: Self-Esteem, Cosmetics, and Cancer* (Doctoral dissertation, The Chicago School of Professional Psychology).

Guo, J., Qiu, Y., & Gan, Y. (2020). Workplace incivility and work engagement: The chain mediating effects of perceived insider status, affective organizational commitment and organizational identification. *Current Psychology*, 1-12.

Gouldner, A. W. (1960). The norm of reciprocity: A preliminary statement. *American sociological review*, 161-178.

George, J. M., & Bettenhausen, K. (1990). Understanding prosocial behavior, sales performance, and turnover: A group-level analysis in a service context. *Journal of applied psychology*, *75*(6), 698.

Hui, C., Law, K. S., & Chen, Z. X. (1999). A structural equation model of the effects of negative affectivity, leader-member exchange, and perceived job mobility on in-role and extra-role performance: A Chinese case. *Organizational behavior and human decision processes*, *77*(1), 3-21.

Harrison, D. A., Newman, D. A., & Roth, P. L. (2006). How important are job attitudes? Meta-analytic comparisons of integrative behavioral outcomes and time sequences. *Academy of Management journal*, *49*(2), 305-325.

Jindal, P., & Tej, F. (2016). Being an educator in private vs. public school system: A comparative study on job contentment. *IRA International Journal of Education and Multidisciplinary Studies*, *3*(3), 309-319.

Kumari, S., Jindal, P., & Chand, P. K. (2022). Social Media Networking: As Facilitator of Acculturation. *ECS Transactions*, *107*(1), 9381.

Jewell, L. N. (1985). *Contemporary industrial/organizational psychology*. West Publishing Company.

Krishnan, S. K., & Singh, M. (2010). *Outcomes of intention to quit of Indian IT professionals.* Human Resource Management: Published in Cooperation with the School of Business Administration, The University of Michigan and in Alliance with the Society of Human Resources Management, *49*(3), 421-437.

Karatepe, O. M., Keshavarz, S., & Nejati, S. (2010). Do core self-evaluations mediate the effect of coworker support on work engagement? A study of hotel employees in Iran. *Journal of Hospitality and Tourism management*, *17*(1), 62-71.

Lau, P. Y. Y., McLean, G. N., Lien, B. Y. H., & Hsu, Y. C. (2016). Self-rated and peer-rated organizational citizenship behavior, affective commitment, and intention to leave in a Malaysian context. *Personnel Review*, *45*(3), 569-592.

Locke, E. A., & Dunnette, M. D. (1976). Handbook of industrial and organizational psychology. *The nature and causes of job satisfaction*, *1297*, 1349.

Lam, W., Chen, Z., & Takeuchi, N. (2009). Perceived human resource management practices and intention to leave of employees: the mediating role of organizational citizenship behaviour in a Sino-Japanese joint venture. *The International Journal of Human Resource Management*, *20*(11), 2250-2270.

Mossholder, K. W., Settoon, R. P., & Henagan, S. C. (2005). A relational perspective on turnover: Examining structural, attitudinal, and behavioral predictors. *Academy of Management Journal*, *48*(4), 607-618.

Mita, M., Aarti K. and Ravneeta, D. (2014) Study on Employee Retention and Commitment. *International Journal of Advance Research in Computer Science and Management Studies*, 2, 154-164.

Malatesta, R. M., & Byrne, Z. S. (1997*). The impact of formal and interactional justice on organizational outcomes.* In annual meeting of the Society for Industrial and Organizational Psychology, St. Louis, MO.

Munene, J. C. (1995). 'Not-on-seat': An Investigation of Some Correlates of Organisational Citizenship Behaviour in Nigeria. *Applied psychology*, *44*(2), 111-122.

Maertz Jr, C. P., & Campion, M. A. (1998). 25 years of voluntary turnover research: A Review. *Management Journal, (35)*, 1036-1054.

MacKenzie, S. B., Podsakoff, P. M., & Ahearne, M. (1998). Some possible antecedents and consequences of in-role and extra-role salesperson performance. *Journal of marketing*, *62*(3), 87-98.

McIntosh, N. J. (1991). Identification and investigation of properties of social support. *Journal of Organizational Behavior*, *12*(3), 201-217.

Noah, Y. (2008). A study of worker participation in management decision making within selected establishments in Lagos, Nigeria. *Journal of social sciences*, *17*(1), 31-39.

Nishii, L. H., & Mayer, D. M. (2009). Do inclusive leaders help to reduce turnover in diverse groups? The moderating role of leader–member exchange in the diversity to turnover relationship. *Journal of applied psychology*, *94*(6), 1412.

O'Driscoll, M. P. (2000). Work and family transactions. *Transactions in the mid-life family*, 92-112.

Organ, D. W. (2014). *Organizational citizenship behavior: It's construct clean-up time*. In Oraganizational Citizenship Behavior and Contextual Performance, 85-97. Psychology Press.

Podsakoff, N. P., Whiting, S. W., Podsakoff, P. M., & Blume, B. D. (2009). Individual-and organizational-level consequences of organizational citizenship behaviors: A meta-analysis. *Journal of applied Psychology*, *94*(1), 122.

Paillé, P., Boiral, O., & Chen, Y. (2013). Linking environmental management practices and organizational citizenship behaviour for the environment: a social exchange perspective. *The International Journal of Human Resource Management*, *24*(18), 3552-3575.

Panoch, A. (2001). *"The relationship between diversity and employee retention."* PhD diss., University of Wisconsin-Stout, Retrieved From http://www.uwstout.edu/lib/thesis/2001/2001panocha.pdf.

Paillé, P., & Grima, F. (2011). Citizenship and withdrawal in the workplace: Relationship between organizational citizenship behavior, intention to leave current job and intention to leave the organization. *The journal of social psychology*, *151*(4), 478-493.

Redman, T., & Wilkinson, A. (2001). In search of human resource management. *Contemporary human resource management*, 3-23.

Raub, S. P. (2018). Perceived insider status and job design predict job attitudes and work performance of restaurant employees. *Journal of Hospitality & Tourism Research*, *42*(6), 880-903.

Roberts, K., Kossek, E. E., & Ozeki, C. (1998). Managing the global workforce: Challenges and strategies. *Academy of Management Perspectives*, *12*(4), 93-106.

Settoon, R. P., & Mossholder, K. W. (2002). Relationship quality and relationship context as antecedents of person-and task-focused interpersonal citizenship behavior. *Journal of applied psychology*, *87*(2), 255.

Shankar, V., Kalyanam, K., Setia, P., Golmohammadi, A., Tirunillai, S., Douglass, T., & Waddoups, R. (2021). How technology is changing retail. *Journal of Retailing*, *97*(1), 13-27.

Stamper C. L., Masterson S. S. (2002). Insider or outsider? How employee perceptions of insider status affect their work behavior. *Journal of Organizational Behavior*, 23, 875-894.

Susskind, A. M., Kacmar, K. M., & Borchgrevink, C. P. (2003). Customer service providers' attitudes relating to customer service and customer satisfaction in the customer-server exchange. *Journal of applied psychology*, *88*(1), 179.

Saraih, U. N., Aris, A. Z. Z., Karim, K. M., Samah, I. H. A., Sa'aban, S., & Mutalib, S. A. (2017). Relationships between organizational commitment, OCB, organizational justice and turnover intention: Evidence from educational institution in Malaysia. *Review of Integrative Business and Economics Research*, *6*(2), 64.

Spector, P. E., Bauer, J. A., & Fox, S. (2010). Measurement artifacts in the assessment of counterproductive work behavior and organizational citizenship behavior: Do we know what we think we know?. *Journal of Applied Psychology*, *95*(4), 781.

Stauss, B., Chojnacki, K., Decker, A., & Hoffmann, F. (2001). Retention effects of a customer club. *International journal of service industry management*.

Smith, C. A., Organ, D. W., & Near, J. P. (1983). Organizational citizenship behavior: Its nature and antecedents. *Journal of applied psychology*, *68*(4), 653.

Stamper, C. L., & Masterson, S. S. (2002). Insider or outsider? How employee perceptions of insider status affect their work behavior. *Journal of Organizational Behavior: The International Journal of Industrial, Occupational and Organizational Psychology and Behavior*, *23*(8), 875-894.

Sarah, K., Oceane, S., Emily, F., & Carole, F. (2021). Learning from lockdown-Assessing the positive and negative experiences, and coping strategies of researchers during the COVID-19 pandemic. *Applied Animal Behaviour Science*, *236*, 105269.

Tang, A. D., Luu, T. T., Chen, W. K., & Liu, S. C. (2023). Internal corporate social responsibility and customer-oriented organizational citizenship behavior: the mediating roles of job satisfaction, work-family facilitation, life satisfaction, and the moderating role of organizational tenure. *Journal of Sustainable Tourism*, 1-21.

Tews, I., Krüger, T., Hebeler, K., & Schwenk, A. (2013). Neutron matter at next-to-next-to-next-to-leading order in chiral effective field theory. *Physical review letters*, *110*(3), 032504.

Thau, S., Bennett, R. J., Stahlberg, D., & Werner, J. M. (2004). Why should I be generous when I have valued and accessible alternatives? Alternative exchange partners and

OCB. *Journal of Organizational Behavior: The International Journal of Industrial, Occupational and Organizational Psychology and Behavior*, *25*(5), 607-626.

Workforce Planning for Wisconsin State Government (2005) Employee Retention. Retrieved from http://workforceplanning.wi.gov/category.asp?linkcatid=15&linkid=18.

Walker, J. W. (2001). Zero defections? *Human Resource Planning*, 24(1), 6–8.

Wayne, S. J., Liden, R. C., Graf, I. K., & Ferris, G. R. (1997). The role of upward influence tactics in human resource decisions. *Personnel Psychology*, *50*(4), 979-1006.

Zineldin, M. (2006). The royalty of loyalty: CRM, quality and retention. *Journal of consumer marketing*, *23*(7), 430-437.

Zhou, J., & George, J. M. (2001). When job dissatisfaction leads to creativity: Encouraging the expression of voice. *Academy of Management journal*, *44*(4), 682-696.

About the Editors

Priya Jindal is currently working as an Associate Professor at Chitkara Business School, Chitkara University, Punjab, India. She earned her doctorate in Management from Maharishi Markandsheware University, Mullana, India. She got a distinction in M.Phil from Maharishi Markandeshwar University, Mullana in 2009. Dr. Priya has contributed her Approx. 16 years in teaching. She supervised four Ph.D. research scholars and two M.Phil candidates. There are numerous research papers to her credit in leading journals among them eleven research paper has been published in Scopus Indexed Journal. She has filed more than 14 design patents and 6 utility patents. Dr. Jindal applied one copyright titled *"Model of 7C's of Service Marketing Mix for Banking 4.0"*. She has an editor of two edited books under the IGI Global Publication. Both the edited books got indexed in Scopus. She has also presented her work at various national and international conferences and received best paper awards for her presentations. She also chaired many national and international conferences including IEEE.

Amandeep Singh holds a Doctorate in Management specializing in Marketing and he is also UGC-NET qualified. He holds more than 16 years of teaching experience. His main area of research is Consumer Sciences and Business Innovations. Currently, he is working as Professor at Chitkara Business School, Chitkara University, Punjab, India. He has published 48 research papers in various journals and conferences which are indexed in Scopus, Web of Science, and Google Scholar. He has edited 7 books published by IGI Global, De Gruyter, and Wiley. He has also chaired many National and International Conferences. He is on the editorial board of 3 International Journals. He was awarded the Best Teacher in 2008. He is also part of the Board of Studies of various B-Schools and leading universities in Northern India.

Rashmi Aggarwal is a Professor and Dean Commerce UG Programs at Chitkara Business School, Chitkara University, Punjab, India. She is PhD, M. Phil, M.Com, and UGC-NET qualified, with a vast experience of 23 years in teaching and administration at various positions. She has also worked at the position of Dy. Dean Students' Welfare handling all the activities related with the students at the University level and outside. She has publications in many reputable journals and has excellent research orientation. She also has many patents and copyrights to her credit. She has organised and attended many FDP's and workshops at University level. She is also an effective communicator with the exceptional relationship management with ability to relate to people at any level. She has the ability in leading and motivating large teams with the belief that smart work and not just hard work leads to success. She has been awarded with the Excellence Award for Most Collaborating Employee and for filing the maximum number of patents. She also got the letters of appreciation for strenuous and meticulous efforts by her in guiding and mentoring the students.

Dr. Rashmi Aggarwal is bringing in much required practical approach and creating a positive learning environment in the classroom. She is an able mentor of the students with excellent feedback from them. She is also guiding the research scholars in the research domain of Accounting & Finance.

About the Contributors

Renjith V Ravi is presently employed as Associate Professor and Head of the Department of Electronics and Communication Engineering and Coordinator of the Post Graduate Programmes conducting at MEA Engineering College, Kerala, India. He possesses B.Tech. degree in Electronics and Communication Engineering in, ME degree in Embedded System Technology and Ph.D. in Electronics and Communication Engineering. He is a member of the panel of academic auditors of APJ Abdul Kalam Technological University, Kerala and had conducted external academic auditing in various affiliated institutions under the same University. He had published several research articles in SCIE and Scopus-indexed journals, Edited books and international conferences inside and outside the country. He is an academy graduate and academy mentor in Web of Science and a certified peer reviewer from Elsevier Academy. He has been serving as a reviewer for various SCIE and Scopus-indexed journals from IEEE, ACM, Springer, Elsevier, Taylor & Francis, IET, Inderscience, World Scientific, IOS Press De-Gruyter and IGI Global. He has been published four edited books and currently editing three edited books from renowned international publishers. He got granted one patent, one industrial design and two copyrights. He had been awarded several outstanding achievement and outstanding service awards, and several best paper awards from international Conferences. He is a Fellow of IETE and a member of IE, ISTE, CRSI, IACSIT, IAENG, SDIWC and a senior member of SCIEI and SAISE, and a chartered engineer certified by the Institution of Engineers (India). He has been served as the Program Committee member, Session Chair as well as reviewer of several National and International conferences conducted in India and abroad.

Govind P. Shinde is Director and Professor of Marketing & Finance at HK Institute of Management Studies and Research, Mumbai. He has a Ph.D. in Management Science, M. Com in Costing and Taxation, MBA in Marketing Management, and also has done Government Diploma in Co-Operation and

Accountancy. He has a rich academic and administrative experience of 24 years. He has worked as I/c Director and I/c Principal in MGMCCSIT Navi Mumbai and has served the organization for 9 years. He has worked in Bharati Vidyapeeth's IMSR, Navi Mumbai for 12 years and has taught the courses like MMS, MMM, MFM, MHRDM, BBA and BCA. He has presented and published more than 58 research papers in various national and international conferences, seminars and reputed and referred national and international Journals, Scopus, Web of Science, UGC Care listed Journals. He is a Research Guide and has guided 6 Ph.D students in Finance, Marketing specializations and General Management, under his able guidance 4 students have got Ph.D. awards. He has published 6 Books related to Management Stream. He has been awarded by Bharati Vidyapeeth Pune for his achievement in academics during the year 2010 Excellence in teaching award, by IEDRA Delhi for excellence in Indian education in 2012 and by ISCM Maharashtra as best professor in Management during the year 2013. He is on the editorial board of various reputed and referred national and international Journals. He is also a member of Various Professional associations in commerce and Management. His few research papers have got best papers awards in conferences. He is subject expert for the University CAS and faculty recruitment in Mumbai University. He is on various bodies of University of Mumbai as a Member, Member of the board of Studies Swami Tamanand Teerth Marathwada University, Nanded.

Roshini T V is presently employed as Professor of Electronics and Communication Engineering at Vimal Jyothi Engineering College, Kannur, Kerala, India. She possesses B.Tech. degree in Electronics and Communication Engineering and M.Tech. degree in Electronic Design and Technology and Ph.D. in Electronics and Communication Engineering. She is a member of the panel of academic auditors of APJ Abdul Kalam Technological University, Kerala and had conducted external academic auditing in various affiliated institutions under the same University. She had published several research articles in SCIE and Scopus indexed journals. She is a Fellow of IETE and member of ISTE, and senior member of IEEE. Her research areas include Image Processing, Machine Learning, Internet of Things Etc.

Saw Mon Yee Aung is employed as an Associate Professor in the Department of Information Technology Engineering Technological University (Toungoo), Myanmar. She received PhD degree from Mandalay Technology University,

Myanmar, in December 2017. She underwent six-month research with the scholarship of the Asian India Research Fellowship Program (AIRTF) in India in 2020. She participated as a member of the editorial board of IGI Global in 2023. She was the external examiner of Ph D candidates at VELS University, Chennai, India, in 2022 and 2023. She published articles in the international research journal in 2022 and the university journal, Journal of Science and Engineering (TUTJSE) in 2020 and 2022 at Technological University (Toungoo).

Suruchi Sharma has a rich experience of more than 20 years in the field of academics, research and administration. She is a proud alumnus of Banasthali Vidyapith (Newai, Rajasthan) qualifying her MBA (Finance and Marketing) and Doctoral Degree in Management from this esteemed organization. She has authored many book chapters, management syllabi of various Government Universities, and presented many papers in national and international conferences. She also has the distinction of having many patents and copyrights in her name. She has been instrumental in leading the faculty team and students towards a number of developmental activities and programs during her tenure as an administrative head. Currently, she is working as Associate Professor, School of Management, Graphic Era Hill University, Dehradun (Uttarakhand, India).

Nidhi Chaturvedi has rich, vast, and proven experience of more than 20 years in the field of Business Management Education. She has completed her MBA & PhD from the very renowned and prestigious institution; Banasthali Vidyapith. Before relocating to Dubai in 2007, she has worked with reputed organizations in India such as Banasthali Vidyapith, IPEM & IMS Ghaziabad in the capacity as Senior Lecturer. Currently she is serving as Program Coordinator and Senior Lecturer 8n School of Business Management at Emirates Aviation University, Dubai, UAE since 2008. Her research areas are Marketing, Consumer Behaviour, Strategic Management and HRM. She is actively involved in reviewing books' chapter, conferences, and research publications in Scopus Indexed and ABDC journals, among others. For her continuous dedication and achievements, she was awarded Certificate of Excellence by Emirates Aviation University, Dubai.

Madhavi Shamkuwar has more than 15 years' experience in Academia. She had authored 3 Text books and authored 6 International book, book chapters (Scopus indexed) and 9 research paper in reputed journals. To her credit their are more than 100 citations for her research articles published. She is pursuing PhD from Department of Management Sciences (PUMBA), Savitribai Phule Pune University. Currently she is working as an IQAC coordinator and is proficient to handle academic and administrative initiatives and activities such as quality initiatives for NAAC and NBA. She is currently working as a Campus to Corporate Training head facilitating the bridge the gap between academic and industry. She has participated in various research conferences, FDP's, Workshops and Seminars. She has coordinated research conferences, FDP's and Seminars.

Pandurang Patil has more than 15 years' experience in Academia and 2 years in Industry. He has completed his MBA (HR and Marketing). He has published 3 research papers at, with 1 paper in ABDC. He is working as Training and Placement-Head. He has participated in various research conferences, FDP's, Workshops, and Seminars.

Rahul Pralhadrao More (M.Com, MBA, Ph.D, NET) is Alumni of IIM Indore with 14 years of teaching. He is the author of 2 Management Books and published 12 Research papers in the UGC care list, ABCD, and Scopus journal. He always updates his skill and knowledge with faculty Development Programmes and workshops. He has tremendous knowledge due to a double P.G (M.Com, MBA) and is also qualified for NET. He contributed to Research at Swami Ramanand Thirth Maradhawada University, Nanded, with Ph.D.

Sujata Barde has 12 years of experience in Academics. She published 2 research papers in reputed journals. She is pursuing a Ph.D. from Savitribai Phule Pune University. She has participated in various research conferences, FDP's, Workshops, and Seminars.

Deepa Sharma has done Ph.D., MBA in Human Resource Management, currently working as Assistant Professor at MMIM, Maharishi Markandeshwar (Deemed to be University), NAAC Accredited Grade 'A++' University, Mullana, Ambala (Haryana). She has more than 7 years of experience in academics. She has written 45+ research papers for various Scopus and UGC listed referred National & International Journals.

Ram Singh is a new Age Innovative Educationist working as a Professor at MMIM, Maharishi Markandeshwar (Deemed to be University), NAAC Accredited Grade 'A++' University, Mullana, Ambala (Haryana). Dr. Singh has vast teaching and research experience of more than 14 years in the UG & PG courses. Being a researcher, Dr. Singh holds multifarious positions in the field of academics.

Sunayna Khurana has done Ph.D., MBA in Marketing, UGC NET qualified, currently working as Professor & Director at MMIM, Maharishi Markandeshwar (Deemed to be University), NAAC Accredited Grade 'A++' University, Mullana, Ambala (Haryana). She has more than 15 years of experience in academics and 3 years of experience in the corporate sector. She has written 55+ research papers for various ABDC, SCI, Scopus and UGC listed referred National & International Journals including International Journal.

Gagandeep is a faculty at Chitkara Business School, Chitkara University, Punjab, India. He received his MBA in Marketing and Ph.D. in Marketing & International Business from Panjab University's University Business School. He has more than 14 years of teaching and research experience. His area of interest in teaching are Marketing Research, Research Methodology, Management Science, Applied Operations Research, Marketing Management and Strategic Management. Dr. Gagandeep has presented papers in various National and International conferences and has published more than 10 research papers. His area of research interest is Empirical modeling of marketing problems, Service Quality, New product development and Emerging trends and issues in marketing.

Jyoti Verma is a faculty at Chitkara Business School, Chitkara University, Punjab, India. With more than 20 research papers to her credit published in various journals and conferences at national & international level, Dr. Verma has developed her core expertise in the area of service quality, blockchain, and behavioral finance. She has authored 1 book titled 'Accelerating the Development of Quality of IT-Enabled Banking Services' by international publishers and published many book chapters in edited books at international level. She has published many patents and has prepared and delivered many E-Content programs on Research Methodology (MHRD Projects).

Kajal Rani presently serves as an assistant professor at Chandigarh vidya jyoti eduversity, Chandigarh; she has four years of academic experience. She has published numerous research articles in UGC care-listed, peer-reviewed journals with a high impact factor. Number of national and international conference attended. She has qualified the UGC NET with high score. She assisted many undergraduates with their research assignments. She has made a significant contribution to the field of business and management.

Natasha Pallavi is an educational professional serving for almost a decade in Education Industry. Currently serving as a faculty in the ICFAI University, Dehradun. Her areas of interest and engagement extend in training students on Professional communication and Life Skills, bridging the gap between campus to corporate life and organizing events at University level. Being an adventure enthusiast and mentor I believe in training and grooming students to be life ready and to face challenges with courage, consistency and determination.

Nipuna Sankalpa Thalpage is a highly accomplished individual who recently completed his MSc in IT from Cardiff Metropolitan University, UK, with outstanding grades.With research expertise in Cybersecurity, Machine Learning, Emerging Technologies and Programming. As a prospective Ph.D. student, he actively participates in conferences and seminars, showcasing his dedication to research. Nipuna's expertise and passion position him to make significant contributions in the field of IT and drive advancements in technology.

Rohit Bansal is working as Associate Professor in Department of Management Studies in Vaish College of Engineering, Rohtak. He has authored & edited 24 books with renowned national & international publishers including IGI Global USA, Scrivener-Wiley Publishing, De Gruyter and Scholar's Press, Germany, etc. In addition, he has published 140 research papers and chapters in journals of repute including Scopus indexed as well as edited books. He has also presented papers in 50 conferences and seminars. His area of interest includes organizational behaviour, marketing management, human resource management, sales and distribution management and entrepreneurship. He is on editorial advisory board as a member in 110 national and international peer-reviewed journals. He served as member of advisory committee in many international conferences. He has acted as session chair in many international conferences. He has been awarded many times for his contribution to academics and research.

Joy Mukhopadhyay pursued his Ph.D. from IISc, India in 1979. A veteran educationist and management consultant with a combined forty-seven years of experience in research, corporate, and academia, he is also a visiting faculty member for several institutes across India. He has also been the Academic Supervisor for Warwick University, UK. Having authored close to a hundred research papers and chapters in journals and books of Indian and international repute, he is on the editorial board of a few journals. He delivers lectures and conducts FDPs, and workshops, both virtually and physically in India as well as globally.

Arka Ghosh is a B.Tech. (Mechanical Engineering) from SRM University, and a gold-medalist MBA (International Business) from ISBR (University of Mysore) in India. A management consultant by profession, he has authored 4 book chapters, 3 conference papers in the management domain, 3 conference papers, and 4 research papers in the mechanical engineering domain. He is a lifetime member of IIIE and IWS and a student ex-member of AIB. He delivers expert talks across institutions in India. Having worked with some huge organizations, his research interest and consulting expertise lie in process engineering, data analytics, corporate strategy, and business transformation.

Ranajit Chatterjee after his Ph.D. from Jadavpur University, India, in 1997, pursued a post-doctoral fellowship from Tokyo Institute of Technology, Japan (1996-1999) in the Department of Computational Intelligent and System Science. He served as a fellow of the Japan Chemical Innovation Institute at the Frontier Collaborative Research Center at Tokyo Institute of Technology (1999-2003) and then as research associate at the University of Electro-Communications, Japan. Since 2004, he is affiliated with the International Rescue System Institute and is currently working as technician-in-charge of the Human-Centered Interface Development project, supported by the JST, Japan. His research interest is in robotics.

Lalita Kumari is currently working as Assistant Professor at Amity School of Engineering and Technology (ASET), Amity University, Patna. She has 14+ years of experience in teaching different subjects of computer science at UG and PG level, including 10 years of teaching experience at NIT Agartala, where she was working as an assistant professor in the CSE department. She did a Ph.D. in engineering from NIT Agartala with the broad domain of Image Processing and MTech. in Computer Science & Engineering from C-DAC, Noida. Her area of research includes Machine Learning, Digital Image Processing, Natural Language Processing, Advanced Algorithm.

Chetna Priti has a Ph.D. degree in Management. She is an Associate Professor and Assistant Director of Amity Business School at Amity

University, Patna. She has teaching experience of more than 18 years at UG and PG levels. She has received award for contribution in publication on Rural Marketing in Indian Management Congress. Dr. Chetna Priti is the Founder Mentor & Advisory Committee member of WECS Association. Through this association she is supporting the cause of women empowerment. She is the Executive Member of the Strategic Trainers and Mentors Association. She is also a Corporate Trainer. She has imparted training to the Mid-Level and Senior Level Executives in IOCL, NTPC, SBI, Bank of Baroda etc. She is the recipient of Women Achiever Bihar from Barhta Bihar Badlta Bihar. She has also received Pratibha Samman award on the International Women's Day from NTPC and from Former Deputy CM Ms. Renu Devi.

Nikhil Sharma is working as Assistant Professor and Program Leader in Amity School of Engineering & Technology at Amity University Patna for last five years. He has a wide experience of 14 years of teaching in the field of Mechanical Engineering (Design). He is also a member of Board of Studies at Amity University Patna. He completed his MTech in Industrial Design from Rajiv Ghandhi Technical University in 2015 and is currently pursuing a PhD from NIT Patna. He has vast experience in teaching like Machine Design, Strength of Material, Fluid Mechanics, Engineering Mechanics, Engineering Graphics, and I.C Engine.

Praveen Nayak is Principal of Vivekanand Institute of Management, Kalaburagi, Karnataka, India. He is a Research Scholar of VTU Belagavi pursuing research on CSR for his PhD studies. He teaches General Management, Strategy and Entrepreneurship. He is a freelance consultant and a Resource Person for Government Organizations and Business Associations. He coordinates Blood Donation activities and is also involved in social welfare activities. His other research area is related to infrastructural barriers to growth.

Preeti Patil is Assistant Professor in Management and teaches Marketing, Research Methodology and Strategy. She has been a Resource Person for several Workshops on Research Methods and also in FDP's. She is also an author and has published Research Articles in numerous journals relating to Marketing and Consumer Behavior. She is actively involved in Quality Assurance and Quality Control measures of the Research Centers. She also coordinates Career Guidance and Campus Placement activities.

Akshita Chotia, Assistant Professor, Poornima University, Rajasthan. I have completed my Ph.D entitled "Exploring the Growth and Development of Woman's Self through the detour of journey in selected American Women Travel Writings" in 2021 from Banasthali Vidyapith, Rajasthan. During my Ph.D, I was associated with Bachelors in German Language Program and was awarded with fully funded scholarship to pursue my German course as an Foreign Exchange student from University of Education, Weingarten, Germany. I have done Double Bachelors, B.A. Pass course and B.A. German Language. I have also sharpened my knowledge of Sanskrit language by pursuing Advance Diploma in Sanskrit. I have had Scopus publications and Book chapters in several Books and have delivered many talks as a Guest Speaker as well.

Sneha Thakur is an assistant professor in the department of English and soft skills at Poornima university. She has graduated with an honors degree in English literature, has been a gold medalist in her MA English, and also holds a diploma in broadcast journalism and trauma counselling. She has submitted her doctoral thesis as a research scholar in Banasthali Vidyapith. Her research interests primarily include the genres of literature- Diaspora and feminist revision mythology, Turkish and afghani literature. However, she has developed keen interest in the area of soft skill training during her teacher career. Her attitude towards research doesn't let her limit herself to literature; she carries an open attitude towards the field of soft skills and education as well.

Garima Bansal is presently working as Head of Commerce Department and Assistant Professor in Hindu Girls College, Sonipat and is having more than 9 years of teaching experience. She has obtained her Ph.D. in Finance from Department of Management Studies, Deenbandhu Chhotu Ram University of Science & Technology, Murthal, Sonipat, Haryana. She did her M.Com from Department of Commerce, Delhi University and B.Com (H) from SRCC, Delhi University. She has authored and co-authored 5 research papers and attended more than 15 seminars of national and international level. Her publications also include one book on Financial Management (Theory and Practice). Her research areas include Financial Management, Financial Well-Being, etc.

Koshish Jyoti is a Research Scholar pursuing her Ph.D in management, with a specialization in HR, OB and finance with school of studies in management,

Jiwaji university, Gwalior. She has a masters in business administration, and specializes in Human resource management .She has certain publications in UGC care listed journals. She was the gold medalist in MBA (FT) at University level. She is a lifetime member of All India Accounting Conference. She is currently working as a Visiting faculty for School of studies in management, Jiwaji University and deal with several technical subjects like stats, operations and so on.

Garima Mathur is Professor (HR & OB), Coordinator MBA Program & HR Specialization with PhD, UGC NET, (Psych) and MBA. With more than 18 years academic experience and an extensive research orientation she has publications in refereed journals including Scopus, WoS and ABDC and guiding scholars for PhD. She represented the institute in inland and overseas conferences. She is a corporate & Academic trainer and has membership of many professional bodies. She is the editor of a journal and authored many books.

Professor Yogesh Upadhyay, Dean of school of studies in management, Jiwaji university, at present he is working in the area of employee engagement, positive psychology and innovative work behaviour/employees' creativity. Additionally, he works on consumer behaviour in the area of hospitality, (Ex-Vice Chancellor ITM University, Gwalior (MP), (Winner-Best Business Academic of the Year Award – 2007 at 60th Diamond Jubilee All India Commerce Conference.

Index

N

O

P

R

S

T

U

V

W